Study and Communication Skills for the Biosciences

THIRD EDITION

Stuart Johnson

and

Jon Scott

OXFORD
UNIVERSITY PRESS

OXFORD

UNIVERSITY PRESS

Great Clarendon Street, Oxford, OX2 6DP,
United Kingdom

2021001006

Oxford University Press is a department of the University of Oxford.
It furthers the University's objective of excellence in research, scholarship,
and education by publishing worldwide. Oxford is a registered trade mark of
Oxford University Press in the UK and in certain other countries

First edition 2009
Second edition 2014
Impression: 3

Published in the United States of America by Oxford University Press
198 Madison Avenue, New York, NY 10016, United States of America

British Library Cataloguing in Publication Data

Data available

Library of Congress Control Number: 2018959892

ISBN 978-0-19-879146-1

Printed and bound in the UK
by CPI Group (UK) Ltd, Croydon, CR0 4YY

Study and Communication Skills for the Biosciences

Preface

The transition from school or college to university is a very significant life event: for all students, it marks the beginning of the next stage in their educational career and, for many, it may be their first experience of living away from home for any significant length of time. This transition, therefore, is associated with a whole range of new experiences in social and academic terms. Among these is a marked change in learning and teaching styles, with an expectation of increasing independence as a learner, and the development of a more mature, critical approach to the subject. Meeting the challenges posed by these changes requires the development of new skills in many areas of life, not least in the areas of study and communication. If you can develop effective study and communication skills early in your university career, you will facilitate your overall learning, and help maximize your academic performance.

This book arose from nearly ten years of experience in developing and delivering a first-year module in Study and Communication Skills for the students in the School of Biological Sciences at the University of Leicester. The structure of the book and its format are based on our experiences of delivering this course, in terms of recognizing where students experience difficulty in developing these skills and in learning from those students better ways of teaching and providing support for their learning. We hope that future generations of students will be able to benefit from this guidance and we also look forward to learning more about the ways in which skills can best be developed.

S.J. and J.S.
University of Bristol and University of Leicester

Acknowledgements

We would like to thank the students in the School of Biological Sciences at the University of Leicester who have helped to shape our work and, over the years, have experienced the evolution of the first-year module in Study and Communication Skills. Thanks are also due to Lucy Wells and Jonathan Crowe from Oxford University Press for their support throughout the project.

Contents

Why are study and communication skills important?

 Introduction

As the title suggests, this book explores the key study and communication skills that are needed by bioscience students beyond school level. This means that if you are studying, or are about to study, biosciences, this book can help. As you may have already gleaned from the contents pages, this book addresses a range of important topics, including:

- using feedback;
- making the most of lectures;
- working with different information sources;
- academic integrity and avoiding plagiarism;
- choosing the right writing style;
- writing essays;
- writing practical and project reports;
- working in tutorials and groups;
- preparing scientific presentations;
- delivering scientific presentations;
- creating academic posters;
- getting the most out of revision;
- getting the most out of exams;
- making yourself employable.

This book has been written to help you in all of these areas and you may wish to work through them one by one. Alternatively, however, it may be that you have a specific issue that you need help with and so may want to skip directly to the relevant chapter. Either approach is fine—this book has been written so that it makes sense when read as a whole text or as individual chapters. However, it is a good idea to read this chapter first as it helps to set the scene.

Good study and communication skills are vital if you are going to do well at university; they will help you to study efficiently and effectively, and so perform to the best of your ability in your chosen area of the biosciences. However, good study and communication skills are not only useful for studying for your course; the skills you need to develop to do well in your chosen field of study have many parallels with the skills that will be useful throughout your career. Being able to research a subject, construct an argument, write a report, present information, manage your time, and plan your own development are all skills that are highly sought after by most, if not all, employers. We hope this book will not just help you to do better in your degree, but will also help you to do well beyond university. The final chapter, 'Making yourself employable', focuses particularly on this.

Incidentally, in this chapter we talk mainly about studying for a degree at university. However, this book is equally useful if you are doing another kind of bioscience course, whether at university or college.

1.1 'I did study skills at school. Why do I still need to read this book?'

Clearly, if you have got as far as getting a place at university, you must be good at studying. Also, you have probably been taught about study skills at school or college, and you will have certainly developed your own study skills during your time there. So why do you need to read this book? The main reason is this—studying for a degree at university is different from studying at pre-university level. There are numerous ways in which it is different. The important ones are as follows.

- Studying at university requires you to be more independent than you were at school. You will, therefore, be expected to undertake much more self-directed study and so will need to learn to *manage your time well*.

- Studying at university requires you to be more critical about your subject than you were at school, and not to just accept everything at face value without question. You will be expected to identify differing views on a topic and develop your own views, too. So you will need to learn to *think critically*.

- Studying at university requires you to be more self-reflective about your ability and performance than you were at school in order to adapt and develop your current skills to better suit an undergraduate level of study. This means reflecting on your work and engaging with feedback to improve what you do in the future. So you will need to learn to *develop yourself*.

- Studying at university requires you to be more focused on future employment. You can't be in formal, full-time education indefinitely—someday you will need to finish your education and start earning money. So you will need to learn to *make yourself employable*.

Studying at university is, for most, a more challenging experience than school. This book will help you to meet those challenges and assist you in making the transition to becoming an accomplished undergraduate student. This chapter focuses on the differences, to highlight what we consider to be the foundational skills you will need to develop (and keep in mind) throughout your studies.

1.2 The contextual nature of skills: making the transition to undergraduate study

Skills are contextual. By this, we mean that skills are learnt in a particular context and the way you apply them is specific to that context. For example, when you learn to drive a car you learn in a certain way—usually with a driving instructor sitting next to you, in a car that's familiar to you, and receiving instruction on what to do. However, just because someone has learnt to drive, it does not mean that person is capable of driving well, in any and every situation. The chances are, initially at least, that the driving skills will be limited to a similar context to that within which they were originally learnt. Read through the following scenario to see what we mean.

- Imagine a friend of yours has spent the past few months learning to drive and they have just passed their test, passing with only two minor faults.
- If they offered you a lift, in a similar car to the one they had learnt in, would you feel safe accepting a lift with them?
- If they offered you a lift, in a similar car to the one they had learnt in, but it was night-time, would you feel safe accepting a lift with them?
- If they offered you a lift, in a similar car to the one they had learnt in, but they were going to drive on a busy motorway, would you feel safe accepting a lift with them?
- If they offered you a lift, but they were driving an articulated lorry (!), would you feel safe accepting a lift with them?

The chances are you would probably be more reluctant to accept a lift the more unfamiliar to the driver the context became. This is because the more unfamiliar a context is to someone the more difficult it is to perform well in that context.

There are clear parallels between this example and making the transition to undergraduate study, which we are sure you will have spotted. You already have study and communication skills, probably ones that have served you well so far, but as we have seen, studying for a degree at university is different from studying at the pre-university level. Your skills will, therefore, need adapting and developing to fit the new situations you encounter. The good news is that skills can be transferred from one context to another, so the skills that you have developed so far are by no means wasted. In fact, quite the opposite; they will provide a very useful starting point.

1.3 Foundational skills

As we mentioned earlier, the important differences between studying at pre-university and at undergraduate level reveal three foundational skills that are vital for students who want to perform well during their degree—managing your time, thinking critically, and developing yourself. These are not the only skills—subsequent chapters will highlight many others—but they do underpin the ones that follow and so are worth drawing attention to at this early stage.

1.3.1 Manage your time well

Many students, on nearing the end of their degree course, wish that they could have their time again. This is often because they feel that, given another chance, they could do better. The reasons for thinking they could do better may be quite diverse, but common to many of them will be the notion that if they had managed their time more effectively they would have been able to perform more effectively and get a better degree. Therefore, managing your time well is crucial to performing well at university. There will be many demands on your time—social as well as academic—and you will have more autonomy about how you choose to spend your time compared with school or college. It is important, therefore, that you learn to allocate your time appropriately. This is one of those things that is easier to say than actually to do; in theory it's straightforward, in practice it's difficult.

We have identified the need to manage your time well as a foundational skill because it is necessary for performing well in a wide range of academic tasks, but that does not mean that we are going to deal with it in an abstract way—with a chapter just about time management, for example. Instead we will address it in what we think is a much more helpful way—giving you advice about how to manage your time, wherever it is relevant, in context: for instance, when writing an essay, giving a presentation, or preparing for exams. As such, effective time management is a theme that occurs throughout the book, so look out for these sections in the chapters that follow.

1.3.2 Think critically

The second foundational skill that we want to draw your attention to in this opening chapter is the need to learn to think critically. Critical thinking is common to all academic endeavours and so it is important to understand what it is. Many people think of 'critical' as a negative term, and perhaps words like 'unfavourable', 'fault-finding', or even 'unkind' spring to mind. Being critical, however, is much more than this. If you have ever read a review of a film, book, play, or concert, the review will have been written by a 'critic'—someone whose professional job is to be critical in its full sense. For example, the critique of the performance of a play comments on all sorts of aspects of that play, including the quality of the performances of the individual actors, the way the stage was designed, the lighting, and so on. Thus, the critic may have written in glowing terms about the leading lady, but felt that the supporting cast was not up to the job. In this context, being critical may mean being very positive and negative simultaneously about different aspects of the same thing. However, it is important to note that this kind of criticism is subjective—one critic's view of the play may be very different from another's.

In the academic sense, the word 'critical' also does not have specifically negative connotations. In fact, in an academic context, being critical is a good thing, something to be encouraged. However, there is one key difference between thinking critically in scientific terms and in terms of reviewing a play—in science, the critical thinking should be undertaken *objectively*. For example, if you are comparing two theories, you should compare the ideas being presented and evaluate the weight of evidence supporting them, using this evaluation to decide which is most tenable. You should approach this evaluation from a neutral (impartial) viewpoint—not looking for something to be right or wrong because you *think* it should be, but because the available evidence indicates it to be so. Critical thinking is, therefore, a process that requires the evaluation of evidence to come to a conclusion that is supported by the outcome of experiments or observations. It is an intellectual exercise that is fundamental to the way in which science works and develops—the basis of scientific method.

As with managing your time, we haven't written a chapter specifically about thinking critically. Rather we will address what it means to think critically when undertaking particular academic tasks—for example, being involved in lectures or tutorials, or when researching information for an essay or presentation. Look out for these sections in the chapters that follow.

1.3.3 Develop yourself

The third and final foundational skill that we have identified is the need to learn to develop yourself. This will not be an unusual concept to you, as doubtless you will have been encouraged to engage in some kind of personal development planning at school or college. To make the transition from pre-university to undergraduate level, however, you will need to continue to develop yourself. Most universities will have a personal development planning scheme in which you will be encouraged to take part. Such schemes are designed to help you in two ways—to improve your academic performance, and to help you make plans for what you will do once you have graduated. It is a process of continuous improvement—thinking about and reflecting on what you have done in the past, specifically thinking seriously about your progress, and engaging actively with the feedback you have received from your tutors and your peers. From these activities you can learn and influence what you do in the future in a positive way (asking yourself questions such as 'What worked well?', 'What didn't go as planned?', 'How can I make sure that the things that didn't work well work better next time?').

Regardless of the format of the system you use, the important thing is that you reflect on your progress (to identify where you are doing well and where you need to improve) and make plans for your future development. You will have numerous opportunities for reflection, including:

- feedback on coursework;
- feedback on exams;
- conversations with your lecturers;
- conversations with your personal tutor;
- conversations with friends;
- time by yourself thinking.

It is important that you use these opportunities to assess how you are doing and then (just as importantly) decide what you are going to do about it. If you have done well on something, then celebrate it in some way (you probably won't need much encouragement to do this!), but also make sure you are aware of why you got a good mark so you can take that approach again. Equally, when you identify areas for improvement, plan what you need to do to improve.

As with *managing your time* and *thinking critically* we have not written a chapter about how to *develop yourself*, not least because the whole book is about developing yourself.

1.4 Make yourself employable

The final element to highlight in this chapter is the importance of making yourself employable. This is not simply one skill, but rather a combination of the whole range of skills that will be dealt with throughout the book. Not only is making yourself employable about *many* skills, rather than just *one* skill, it is also about experience, too. It is also about the ability to articulate these skills and experiences, in a clear and convincing way to potential employers. Given the breadth and importance of making yourself employable, this is something that we have written specifically about in Chapter 15, 'Making yourself employable' and we would urge you to read it at an early stage, rather than making the common mistake that it is something you can delay thinking about until your final year. Making yourself employable might not be something that you have at the forefront of your mind when you start at university—but you should! Gone are the days when a degree almost guaranteed you a graduate job. There are now far more graduates looking for jobs, and the current economic climate means that some sectors are particularly challenging. This doesn't mean that a degree isn't worth it (on average, graduates still have a much higher employment rate and earn higher salaries than non-graduates) or that there are not enough graduate jobs out there (some sectors are growing, rather than shrinking). However, it does mean that finding a graduate job is much more competitive than it used to be; you will need to work harder and start earlier at making yourself stand out from the crowd.

 Chapter summary

Good study and communication skills are vital if you are going to do well at university and in future employment. Undergraduate study is different from pre-university study in a number of important ways. The skills you have already gained up to this point will be useful, but you will need to develop and adapt them for the new context you now find yourself in. Managing your time well, thinking critically, and being committed to self-development underpin the other skills that we will now explore in the subsequent chapters. These, together with appropriate experience and the ability to articulate your strengths to potential employers, are what will help you not just to succeed in your studies, but to succeed in your working life beyond university.

2 Using feedback

Introduction

So the essay you have worked so hard on has been marked and returned to you. You take a quick look at the mark—it's quite good and about what you expected, so you put the essay into a folder and don't look at it again. You have just done what many students do when they get a mark that was about what they expected or maybe a little better—you feel quite satisfied, but you have just denied yourself an important opportunity to improve your future performance. You got quite a good mark, but why wasn't it really good? What could you have done better? How can you get a really good mark next time?

This is where feedback, and how you use it, is important, and is a vital component in the foundational skill of developing yourself that we identified in Chapter 1. This also reflects a common difference between your experience at university and that at school or college. You may have been used to drafting a piece of work, receiving feedback on it, and then redrafting it before finally submitting it. At university the common practice is that you prepare the work and submit it for marking; the feedback will accompany the marks. Therefore you need to get into the habit of using the feedback from one piece of work to improve on the next one. It is all too easy, however, to make excuses for not doing anything: 'I'm too busy at the moment, but will look at it when I get the chance'; 'I did the piece of work so long ago, there's no point in going back over it now'; 'We've finished that module, so there won't be anything useful anyway'; 'I never understand/can't read what Dr Bloggs writes, so there's no point in looking at it.' No doubt you can come up with quite a few more seemingly plausible reasons for not bothering to revisit a piece of work, maybe not even bothering to collect it in the first place. However, whatever form the feedback takes, even if it is really limited and you only have the bare marks, you should always make the effort to review the feedback, and make use of it to improve your future work.

2.1 What is feedback?

Ideally, feedback should serve three key functions:

1. provide you with a clear indication of how well you are progressing in your work in relation to the standards established by your university;
2. give you guidance as to how you can improve for when you come to prepare the next piece of work;
3. flag up things you are doing well and should therefore continue doing.

At its most basic level, then, feedback takes the form of the marks you get for the piece of work—these marks give you an overall indication of how well you are doing. In terms of specific guidance, marks on their own are clearly not very helpful; nonetheless, you can still use them to help you improve (see Section 2.3, 'How to make use of feedback').

One level up from this, in terms of usefulness, is feedback that indicates what you got wrong or did badly. This is probably the most common form of feedback and it may take the form of specific comments such as 'you did not discuss the role of . . .' or even 'this is wrong'. The marker may also include more generic comments such as 'the essay was poorly structured' or 'not enough detail'. These types of comment are very common, in part because they are usually the easiest type of comment for a marker to make. By pointing out what is wrong, the marker is providing limited guidance as to areas for improvement, but it is often not clear exactly what you should do to improve, particularly in the case of the generic comments, which may be quite vague. In this context, therefore, the best feedback is that which identifies the error and then provides guidance on how to overcome it. This may take the form of full examples or, perhaps, pointing you in the direction of specific resources that can help you:

> 'Make sure that you use the appropriate referencing style for your essays—look at the section on referencing in the course study guide.'
> 'You have misunderstood the mechanism of oxygen transport—see pages 32–35 in your physiology course text.'

A less common form of feedback is positive feedback. Here, the marker not only provides encouragement: 'this was a good essay', but also identifies specific aspects of the work that were well done:

> 'This was a very good essay and I particularly liked your use of diagrams to support the arguments you were making.'

Again, the value of the feedback is greatly increased where examples are given so that you are clear about what things you should carry on doing. Inevitably, the quality and level of detail of the feedback you receive will vary. Indeed, recent National Student Surveys (NSS) have consistently identified feedback as one of the aspects of higher education that students find least satisfactory. However, it must be remembered that feedback is a two-way process and you will only benefit from it if you engage with the process. In this way, even if the feedback you receive is not always very detailed or is late in coming, you will still be able to identify ways of improving what you do.

2.2 When do you get feedback?

Many students only think of feedback in terms of written comments on coursework assignments. Indeed, this is the most obvious and one of the most common forms of feedback. However, feedback comes in a variety of forms and, unless you are ready to recognize when you are receiving feedback, you will be missing out on valuable learning opportunities. Let us now consider the range of different feedback types that may be available to you.

2.2.1 Written comments on written assignments

As stated earlier, written comments on written assignments constitute the most obvious form of feedback and are probably the most *common* form of formal feedback. These types of comments usually take two forms:

1. generic comments, usually written at the end of the piece of work, or on a separate cover sheet—these comments tend to relate to the overall style and quality of the essay;

2. specific comments, which usually take the form of annotations made on the script itself —these may relate to stylistic issues, such as paragraph structure or referencing, or they may be subject-specific, such as the identification of a factual error or the omission of a point in the argument.

2.2.2 Comments on oral presentations

These feedback comments may be given in the form of written comments on a mark sheet, but they are often given just as verbal feedback following the presentation. Such verbal feedback can be very useful because it is the most immediate form of feedback, being given within a few minutes of the delivery of the piece of work. However, verbal feedback is also transient: if you don't write the comments down, you are likely to forget them very quickly, and then you won't derive any long-term benefit from them.

2.2.3 Comments during tutorials

Tutorials are often a source of very useful discussion about specific topics in the curriculum (see Chapter 9, 'Working in tutorials and groups'). However, as with the verbal feedback on oral presentations, the points made by a tutor will be transient and quickly forgotten unless you note them down. Tutors probably consider that they are providing you with valuable feedback during tutorials, but students often don't recognize it as such.

As well as academic tutorials, you are also likely to have meetings with a personal tutor or adviser to discuss your progress as part of your development planning. These are a good opportunity to talk about the feedback you have received on different pieces of work.

Why is this important? First, it provides a good opportunity for discussing feedback that you have received across several pieces of work. However, it is important to remember that, at university, many markers may only mark one piece of work you have submitted. If they make a comment on your work, for example, about paragraph structure or use of references, they will not know if they are the first person to make that comment or if several other markers have also said the same thing, regarding other pieces of work. Only you will be in a position to bring those different pieces of feedback together, so it is useful to be able to discuss this with a personal tutor or adviser.

Try this—Common feedback comments

Collect together the last three or four pieces of work that you have had marked and read through the feedback. Make a note of any comments that are made more than once. These are clearly areas for you to focus on in future and possibly discuss with your tutor.

2.2.4 Comments during practical classes

Practical classes offer another very useful way in which students can engage in discussion with tutors or demonstrators. As a consequence, they can be a rich source of feedback, for example, in relation to experimental technique, analysis and presentation of results, or in understanding of the topic itself. Again, however, much of this feedback is transient and it is very important that you make a note of the comments before they are forgotten.

2.3 How to make use of feedback

If you are going to benefit from the feedback that is provided, then you must take steps to engage with it as an active process. This doesn't just mean looking briefly at the comments. Engagement requires a detailed reflection of how the comments relate to the piece of work you have submitted and consideration of how you should change your approach to future pieces of work.

There is often a perception that if feedback is not provided very quickly, then it is of no value. The usual reasons for taking this view are:

- that it is several weeks since you did a piece of work and you can no longer remember much about it;
- that the module for which you did the piece of work has now finished and, therefore, the feedback no longer appears relevant.

Both these reasons, while initially understandable, are false. There are always useful pointers that you can take from feedback, although it may be that sometimes more work is required on your part to extract the maximum value.

For the most part in this section, we will focus on making use of written feedback, since this is the most common form of formal feedback on coursework; however, much of the guidance also holds true for verbal feedback. The difference between the two is that, in the latter case, it is up to you to make sure that you make a record of the feedback.

2.3.1 Engaging with written feedback

If you have a cover sheet, as is used by many universities, or if you have been given a summary of written comments at the end of the essay, then read these carefully before doing anything else. Divide the comments into two groups.

1. *Generic comments*: ideally, these will take the form of 'feed forward' comments, such as identification of broad areas for improvement. This could include how to improve your overall writing style, to reference better, or to improve the way you draw graphs and present data. There may also be positive comments, identifying things that you have done particularly well and things you should carry on doing in future work.

2. *Subject-specific comments*: these will normally be related to factual information, such as the identification of factual errors or omissions. Again, markers should also flag up good points, so you get a feel for what you are doing well and for what needs improving.

Having categorized the comments, read through the piece of work carefully—it may be some time since you wrote it. As you read through, bear in mind the comments that were made and try to be objective in your evaluation. Make sure that you can identify what aspects of the essay or report the comments relate to. You should also note any annotations on the script. With the generic comments, make sure you understand how the comments apply to your work and how you can improve what you have done. Make a note for future reference so that you can carry them forward to your next assignment. Although the next assignment may well be for a different module, generic issues such as referencing style or essay structure will still be relevant. Likewise, make a note of aspects of your work that were highlighted as being particularly good. With subject-specific comments, check that you understand any factual errors or omissions. Make a note to go alongside your lecture notes to make sure that the point is clear for revision purposes.

Look at your marks and check them against any marking criteria that you might have been given for the assignment or against any generic criteria that your department might have published. Make sure that you appreciate how your work fits against the criteria and why it falls into the specific mark band that it was allocated to. Again, if you don't understand, then ask.

For example, if you obtained a mark of 55%, the marking scheme might state that the work has the following characteristics:

> The essay is well organized, displaying understanding of the main issues. There may be a few, minor errors or poorly expressed ideas. There is a significant dependence on lecture notes and/or textbook material.

As you read through your essay, make sure you can identify why the marker placed it in this category—what were the minor errors? Did you mainly use lecture notes or textbooks, rather than research literature (see Chapter 4, 'Working with different information sources')? Think about how you could do it better next time.

Feedback may not always be clear! For example, if handwritten, the comments may be illegible, or perhaps you don't understand how to improve, even though the marker has highlighted a weakness. You might not understand what the marker means by some comments. If you are not sure what to do, then do ask your tutor to explain in more detail how you can improve—arrange a meeting with the tutor, and take the work along with you so you can discuss the strengths and weaknesses in detail.

Make sure that you take the feedback forward; use it to improve your next piece of work and your understanding of the subject. When you are preparing for your next assignment, remember to read back through the notes you have made on previous pieces of feedback so they are fresh in your mind. When you have taken the points forward, check the feedback on the next piece of work to ensure that you have, indeed, improved.

Try this—Swap your feedback comments

Try this feedback activity with a friend. Take the most recent piece of work you have had returned and swap the feedback that you received. Identify the skills that your friend needs to improve on and encourage them to note down how they can improve them. Ask your friend to identify the skills that you need to develop and make a note of how you can improve them. Then, make sure that you refer back to these notes before beginning the next assignment.

2.3.2 **Engaging with oral feedback**

As indicated earlier, this may take various forms, for example in the laboratory, a tutorial, or as feedback on a presentation. The danger with any of these forms of feedback is that they are very transient and easily forgotten; therefore, you must *be sure to take notes* so that you can remember the specific points. The principles, however, are the same as for written feedback—divide the feedback up into generic and specific aspects, and make sure that you understand what the issues are and how to take them forward.

Feedback, then, is one of the keystones of good learning, but you cannot treat it as a passive process, otherwise you will not progress, and you will make the same mistakes again and again. Engagement is all-important.

 Chapter summary

Figure 2.1 shows a flow chart of the processes of using feedback. The most important aspect of this is that you need to think about the feedback, see how it relates to your work, then to see how you can use it in future to improve your performance.

Figure 2.1 Engaging with feedback

3 Making the most of lectures

 Introduction

Lectures remain a prominent and important feature of almost all undergraduate bio-science courses; therefore, being able to make the most of them is an important skill. The lectures you have as part of your course will be many and varied; your lecturers will have different styles and different levels of ability in lecturing; and the lecture content itself will be of more or less interest to you, and more or less demanding. Regardless of these variables, however, you need to make the most of lectures. This chapter will help you understand how to make the most of lectures by, first, addressing the purpose of lectures, and then giving advice on how to:

- prepare before lectures;
- listen and make notes during lectures;
- follow up after lectures.

3.1 The purpose of lectures

Lectures can be a very valuable resource as they can synthesize the views of several researchers and textbooks, or provide new, even unpublished, information. Lectures are often liked by academics as a means of communication because they are a very efficient means (in theory at least) of transferring lots of information to large numbers of students all at the same time. How efficient a lecture actually is at communicating information will depend on at least three variables:

- how well the lecturer prepares and delivers the lecture;
- how well the lecturer engages the students throughout the lecture;
- how well the students are listening, making notes, and participating.

Clearly, you cannot do much about the first of these variables (how well the lecturer prepares and delivers the lecture), at least not in the short term. However, do make sure you feed back your thoughts to staff on how you found a particular module, usually via the end-of-module survey. (If you think your feedback needs acting on more urgently, contact the module convener directly.) The third variable (listening, making notes, and participating) is an area you *can* do something about; in fact, no one else can do it for you!

3.1.1 More than information transfer

Lectures are not merely about transferring information from the lecturer to the audience. In fact, we would argue that they are not even primarily about transferring information from the lecturer to the audience. If the purpose of a lecture was only to transfer information from the lecturer's brain to the students' notes, then lectures really have very little point, because there are many more efficient ways of doing this, for example preparing a handout or suggesting particular chapters of a core text to read.

Lectures are of value because they add something in the communication of information that couldn't be achieved just as well by other means. You will know this if you have ever tried to read a handout or a friend's notes from a lecture that you didn't attend; trying to understand the information out of context can often be very difficult. So if lectures aren't primarily about transferring information from the lecturer to the audience, what should they be about? The transfer of information is important, but it is what you do with the information as you receive it, and afterwards, that really matters. This is where you, as a student, come in. It is perfectly possible, even sitting in a lecture that has been well prepared and is being well delivered, to disengage your brain and relegate yourself to being a mere passive recipient of the information.

As you will see later in the chapter, it is possible to not think about a lecture, even while you are taking notes on it. The amount of notes you take is not the measure of whether you have been paying attention; in fact, you might have a lot of notes because you haven't been paying attention! Doing something useful with the information as you receive it depends on processing the information effectively. This processing can't all take place within the lecture itself (even if you do listen actively and make notes appropriately), which is why this chapter covers preparation before lectures and follow-up after lectures, as well as what to do during lectures. Before we address these issues, it will be helpful to consider how lectures are different from what you may have experienced at school or college, prior to your undergraduate degree.

3.1.2 Differences between lectures and lessons

Lectures at undergraduate level will be different from lessons you had at school or college in a number of important ways. At the risk of generalizing, these differences probably include the following.

Class size

In years 12 and 13 at school, you may well have been in a class of less than 30 students; at undergraduate level it's not unusual, especially in the first year, for class sizes to be up to 200–300 students (although typically, as you move through your course to the final year, the class sizes will get significantly smaller).

Anonymity

A bigger class makes individuals within the class more anonymous; it is easier in a larger class just to sit quietly and not really engage with the lecture content because you think no one will notice.

Relationship

Bigger class sizes mean that you are more remote from the lecturer, not only in terms of phys-ical space, but also in terms of relationship. The chances are that, in your first year of study, you probably won't get to know most of your lecturers very well, and they probably won't even know your name.

Completeness

In school, your lessons probably covered everything you needed about your syllabus. Importantly, at undergraduate level, lectures are not intended to tell you everything you need to know and so need to be supplemented with additional reading (but more on this one later).

3.2 Prepare before lectures

You may already be on a busy course with a lot of timetabled teaching hours—made up of lectures, tutorials, and practical classes—so is preparing for lectures really necessary? We think it is. Preparing for lectures doesn't have to take long and can actually save you a lot of time. It doesn't have to take long because you are not trying to cover all the material the lecture is going to cover (that's what the lecture is for). Instead, you are just giving yourself a framework to help you understand it better. It can save you a lot of time, because if you have a frame-work to help you understand the lecture, you can process the information more quickly, thus making the listening easier, the note-making more selective, and the follow-up more focused. Preparing for lectures makes for a much more productive learning experience. In order to know how to prepare for lectures, though, you first need to know some basic information about your course.

3.2.1 Know your course

In this context, by 'know your course', we mean finding out how your lectures relate to the course as a whole. This will involve finding out answers to questions such as:

- Most importantly, what are the learning outcomes for the course; in other words, what should you be able to do by the time you have completed the course?
- If the course includes tutorials, do they prepare you for the lecture or do they follow up on material covered in the lectures?
- If the course includes practicals, do they prepare you for the lectures or do they follow up on material covered in the lectures?
- Are there opportunities to discuss the lectures in tutorials?
- Is there any recommended pre-reading; if so, what is it?

Answers to these questions should be available in your module handbook or, alternatively, from your tutor. Additionally, you will need to know the title of each lecture, or group of lectures, within any given module; this, too, should be in the module handbook, which may also contain further details about what each lecture will cover.

3.2.2 Reading before the lecture

We have already noted that preparing for lectures doesn't have to take long because you are not trying to cover all the material the lecturer is going to cover, but are just giving yourself a framework to help you understand it better. A framework will help you understand the lecture, because it gives you somewhere to place information and so relate it to other information, rather than trying to understand it in isolation (which is always very difficult).

The framework can be quick to create, because you only need to understand basic structure and terminology, rather than detail. The lecturer may upload all the slides or lecture content onto the virtual learning environment (VLE) beforehand; if so, have a look at them and download them so you can annotate them during the lecture. A good way of supplementing this is by looking up the relevant chapter in the core text for the module (this is where you need to know the title of each lecture) and scanning through it to familiarize yourself with the main themes. This doesn't require you to read the whole chapter, or read it in detail, just scan through all or some of the following:

- the chapter overview (if there is one);
- the chapter headings and subheadings;
- the introduction to the chapter;
- the conclusion or summary to the chapter;
- figures, diagrams, or tables (which often summarize a lot of information succinctly).

Also, look out for terminology that you are unfamiliar with and try to find out what it means.

Core texts are not the only source of information for pre-reading, however; your own notes will be useful, too. Assuming you are not on the first lecture of a module, your notes from previous lectures in the module will be an important source of information. Again, you don't have to read all the notes in detail, just scan through them to remind yourself of what was covered previously. If you come to a section that you don't understand, spend more time on it and check up on the topic in a textbook or other reference source, or ask your friends.

If you have prepared for a lecture it makes listening and note-making much easier.

3.2.3 Pre-lecture activities

Your lecturers will want you to arrive at their lecture prepared for what they are about to deliver. Some of them may use formal pre-lecture activities in order to help you prepare. These may take many forms; they may ask you to read a section from the recommended textbook or to read ahead in the lecture notes; some lecturers will adopt a 'flipped' approach, where the lecture itself is made available as a recording beforehand and then they will use the lecture time to focus on some specific points or generate a discussion (see Box 3.1). If you haven't gone through the pre-lecture activities you will quickly find that you are at a significant disadvantage. Some lecturers may start their lecture with a quiz or a few short questions, based on the previous session. All these activities are very useful to engage with and have been

> **BOX 3.1 The flipped lecture**
>
> Many lecturers now audio- or video-record their lectures. This has led to some different approaches to the traditional lecture. One of these is commonly referred to as 'flipped lectures'. As the name suggests the approach is very different to a traditional lecture where the lecturer gives information to the students in the lecture. In a flipped lecture this situation is reversed—you will be expected to have studied the content of the lecture in your own time, often by watching a video lecture or by studying complete handouts. The 'lecture' sessions are then used for students to be engaged in activities such as answering questions, perhaps using audience response systems, or discussing and sharing their knowledge and understanding with other students. You may find this approach daunting at first, as it may be very new to you. However, its popularity is down to the belief that you will learn more effectively this way. So you should make sure that you engage seriously with any pre-work that has been set. Do this actively, and make notes and annotations for yourself (see Section 3.4, 'Make notes appropriately during lectures'). Then, when you arrive in the classroom or lecture hall, make sure that you attempt all of the activities that are given to you. This is where your understanding will be consolidated and where you will have the chance to practise the types of activities that may be required for assessment. Clearly, this is a very different approach to the traditional lecture and many of the lectures that you have will actually fall somewhere between these two extremes, with a mixture of pre-lecture preparation and active engagement during the classroom sessions.

designed explicitly to help you understand what is coming in the lecture. So take them seriously and always do any pre-lecture activities that are prepared for you.

Try this—Preparing for a lecture

Read through the notes you took from the previous lecture and make sure that you understand all the material. If there are any aspects that you don't understand, check them in a textbook or ask your lecturer. Go through any pre-reading and make notes from it. If there isn't any pre-reading, skim-read the topic in a textbook so that you get an overview of what is to be covered.

When you have done this and have attended the lecture, think back on how well you felt able to follow the lecture compared with one where you have gone in unprepared. Try to get into the habit of preparing for every lecture.

3.3 Listen actively during lectures

This section is deliberately titled 'Listen actively' because there is a difference between just listening and listening actively. Listening actively suggests that you are alert, attentive, and ready to engage with the content of the lecture, as opposed to just being there. Clearly, the quality of the lecture—how well it is prepared and delivered—will have an effect on how easy it is for you to listen, but listening actively means that you are willing to work hard at listening, regardless of the quality of the lecture. Additionally, we have already identified that preparing for a lecture makes listening actively much easier, because you have a framework within which to place new information and you will at least recognize the terminology used.

3.3.1 Identify your priority

There will be some lectures when it will be difficult both to listen and to make notes, either because of the complexity or newness of information being communicated, or simply because of the amount. In such situations you will need to identify what your priority is—is it to listen or is it to make notes? Deciding which your priority is will depend on whether you are more concerned with trying to understand the information or collect information. If your priority is *understanding* the information then you will need to:

- focus on listening;
- make only brief key word notes;
- if you have downloaded the slides the notes can be very brief annotations;
- follow up the lecture by making detailed notes.

If, however, your priority is *collecting* information, then you will need to:

- focus on making notes and making detailed annotations on the slides;
- make more detailed notes;
- follow up the lecture by reviewing your understanding of the content.

When making a decision about your priority, also consider what resources are available to you after the lecture. It is usually relatively easy to get hold of appropriate textbooks or perhaps borrow a friend's notes, and you may have access to lecture handouts or, more commonly, an electronic copy of the slides. Many universities provide lecture capture facilities to students; these typically enable you to review the slides and audio, and they may also have the video of the lecture as well. Such facilities can be very helpful if used strategically, either to preview or review content, for example you can go back over a section of the lecture where you had difficulty in keeping up. They can also be very useful for revision. Importantly, however, they are not an excuse for non-attendance! Even if you can listen to a recording, it is rarely as good as actually being there. This would suggest that active listening and, therefore, understanding should be more important than making notes. Remember, however, you do not necessarily have to keep to the same strategy all the way through a given lecture; you could switch between the two, depending on the material (more on this in Section 3.4, 'Make notes appropriately during lectures').

3.3.2 Listen for structure

Making notes is always easier if you have an awareness of the structure of the lecture. Being aware of structure enables you to be more selective and, therefore, more focused, in your note-making because it gives you an indication of what the important bits are. That is not to say that the other bits are unimportant; it's just the recognition that, in terms of understanding, there are certain elements of a lecture that are helpful to grasp in order to understand the rest of it.

Sometimes the structure of a lecture is very clear, other times less so. It is important, therefore, to know what the structural clues might be. The most obvious, and probably the most common way of making the structure of a lecture clear, is for the lecturer to outline what the structure is going to be at the beginning of the lecture, either as a list of head-ings or as a potted summary of the content. If a lecturer does outline the structure, make sure that you make a note of it straight away (this applies equally, whether your priority is focusing on understanding or collecting information), as it will give you a sense of direc-tion and help you to anticipate points or take up the thread of information again, should you get lost.

Additionally, during a lecture, the lecturer will probably give you some cues or 'verbal sign-posts', these include statements such as:

- 'I shall now discuss . . .'
- 'My next point is . . .'
- 'Finally . . .'
- 'In conclusion . . .'
- 'To summarize . . .'

These signposts identify a new point or stage in the lecture, and you should show this in your notes accordingly. Other signposts include pausing to indicate a new point or summarizing what has been said prior to moving on.

There are other, more subtle, verbal signposts, which can help you structure your notes; you will need to listen for these. Examples include:

- 'On the other hand . . .'
- 'Others have argued . . .'
- 'Turning now to . . .'
- 'Alternatively . . .'

Other words and phrases indicate that an illustration is about to be given:

- 'An example of this is . . .'
- 'This can be seen when . . .'
- 'Evidence for this can be found in . . .'

Your ability to listen actively and, in particular, to listen for structure, will improve with experi-ence. As you improve you will be better able to spot digressions or additional examples, and adjust your note-making accordingly, which brings us on to the next section.

3.3.3 Engage with in-lecture activities

Your concentration levels may begin to drop off during a lecture. So, many of your lecturers will build activities or breaks into their lectures to help you refocus and regain your concen-tration. These lecture activities can take many forms, depending on the nature of the topic being taught, the constraints of the lecture room, the size of the class, and the preferences of the lecturer. Whatever these activities are, you will find them refreshing if you take part enthusiastically, rather than use the opportunity to chat about last night's entertainment.

However, even that will help you to regain some focus. During an in-lecture activity you might be asked to:

- discuss something with your neighbour and feed back;
- stand up if you . . .;
- raise a hand if you . . .;
- vote using coloured card or raising your hand;
- watch a video;
- write a question on a post-it note;
- answer multiple-choice questions using an audience response system.

3.4 Make notes appropriately during lectures

We commented earlier (Section 3.1.1, 'More than information transfer') that the number of notes you take is not the measure of whether you have been paying attention; in fact, you might have a lot of notes because you haven't been paying attention. Therefore, we need to draw a distinction between *making* notes and *taking* notes. We have deliberately referred to making notes so far, as opposed to taking notes, because we think there is an important difference.

If you *take* something, you are just the recipient of it, so in the case of lectures you just sit there and write down what the lecturer is saying. On the other hand, if you *make* something you are involved somehow in the creation of it and you bring something of yourself to it, so in the case of lectures you create notes that are unique to you and are produced by you engaging thoughtfully with the content of the lecture. Making notes, therefore, is an active process, whereas taking notes is a more passive process. As you can see, there's quite a difference. That's not to say that making notes won't involve you writing down, word-for-word, what a lecturer is saying on occasions; the important thing is that you make notes appropriately. As we have already seen in Section 3.3.1, 'Identify your priority', this will depend a lot on what your priority is; whether it is to understand the information (and so you focus on listening) or to collect information (and so you focus on note-making). In reality, although it is difficult to focus on both listening and making notes at the same time, it is not that you just do one or the other during any given lecture; rather the two go hand-in-hand. You will be constantly switching between listening and making notes all the way through a lecture, but how much you do of each should reflect your priority, as described in Section 3.3.1.

3.4.1 Know why you are making notes

We have already encouraged you to think about what your priority might be when in lectures, and understanding information and collecting information are two key reasons for making notes. However, there are other reasons, too. Making notes appropriately in lectures can also help you to:

- concentrate better;
- remember the content better (in the short term at least);

- think about questions that you want to ask;
- highlight areas of interest.

Note-making, therefore, isn't just about having something to refer to later; it is part of the learning process itself. This is another reason why borrowing a friend's notes is a poor substitute for actually being there and experiencing the lecture for yourself, because so much takes place in the lecture, and in your head, that even very good notes can't really capture the whole event.

3.4.2 Know how to make notes

The ability to make good notes is a skill that develops with practice, so don't expect to be an expert at it straight away. Clearly, you will have made notes before at school or college, but making notes in lectures in an undergraduate setting is different for a number of reasons, as highlighted in Section 3.1.2, 'Differences between lectures and lessons'. So improving your note-making skills comes partly through practice, but there are some important principles to bear in mind, too. These include:

- using structure in your notes;
- using your own words;
- using fewer words;
- using abbreviations;
- using space;
- using colour and image;
- using handouts or downloads of the slides;
- organizing your notes.

We will address each of these principles in turn.

Use structure in your notes

Notes that are lacking in structure will be much more difficult to understand, especially after the lecture, than notes that have a good structure. Imagine you were looking at some lecture notes a few days after a lecture that looked like the ones in Figure 3.1. Notes like these are difficult to understand, partly because they are lacking in structure.

Alternatively, the notes represented in Figure 3.2 have better structure and so should be much easier to understand. The main purpose of structure is to make clear which information is important. It is crucial to note that identifying important information is much easier during the lecture because lecturers will use their tone of voice, pace, and many other devices to provide emphasis to the material, thus giving you an indication of what aspects of the information are most important. After the lecture, however, it is very difficult to recall this kind of detail.

To give your notes structure, you should:

- use headings to order information;
- give each point a new line;

Figure 3.1 Poorly structured lecture notes

Respiration – Prof Smith   22 Feb

Inspiration – depends on contraction of diaphragm – flattens – and of external intercostals. Cause increase in thoracic vol., reduces pressure, draws air in.
Expiration – passive process – elastic recoil of lung tissue. Air flow is tidal, air in lung is not the same composition as air outside – has more CO_2, less O_2 'cos some remains in lung at end of expiration – even if forced. Inspiration has to do work – overcome elastic tension of lung tissue and surface tension of fluid lining lungs. ?role of surfactant??
Gas exchange – occurs in terminal bronchioles and alveoli. O_2 dissolves in fluid and diffuses across lung wall to blood vessels – down concentration gradient. O_2 poorly soluble so needs large surface area for dissolving. CO_2 is better, goes in opposite direction. Need short diffusion path.

Figure 3.2 Well-structured lecture notes

Respiration – Prof Smith   22 Feb

Inspiration*
Dia. + ext intercostals contract
thoracic vol ≠ + pressure ∅
air drawn in
* has to do work – overcome elastic tension of lung tissue and surface tension of fluid lining lungs. ?role of surfactant??

Expiration
• passive process – elastic recoil of lung tissue
• air flow is tidal
• air in lung is not the same composition as air outside – ≠ CO_2, ∅ O_2 (some remains in lung at end of exp.)

Gas exchange
• occurs in terminal bronchioles + alveoli
• O_2 dissolves in fluid and diffuses across lung wall to blood vessel – down concentration gradient
• O_2 poorly soluble so needs large surface area for dissolving
• CO_2 is better, goes in opposite direction. Need short diffusion path.

- highlight examples and illustrations in an appropriate fashion;
- use diagrams to summarize information;
- make clear when sections of your notes are digressions from the main points.

Use your own words

One of the reasons why it is important to try to use your own words when making notes in lectures is because it will help you (or perhaps force you) to understand the content of the lecture better. If you are trying to put information in your own words, then you must process it as you hear it in order to express it in your own words. It also helps you to make notes, as opposed merely to take notes (as explained at the start of Section 3.4). You may have experienced occasions in lectures when you have been able to write down what the lecturer is saying without actually thinking about what is being said—this illustrates how passive the process of note-taking can actually be. How much you try to put information into your own words will depend, to a certain extent, on your priority, but whether your priority is to understand or collect information, putting that information in your own words will help you understand it better. This does not mean that you have to put absolutely everything in your own words; for example, there are two particular occasions when it is important to record the precise wording:

1. *When you are recording a quotation that the lecturer is referring to*: in this case make this clear in your notes by using quotation marks.

2. *When you don't understand what the lecturer's words mean*: in this case make this clear in your notes by adding a question mark in the margin (for example) as a reminder to follow up the point later.

Know how much to write down

There are two potential problems related to how much information you write down—writing down too much information or not writing down enough. How much you write down will, again, depend on your priority (understanding information or collecting information), but it is also influenced by your own level of confidence. Under-confidence tends to lead to writing down too much, whereas over-confidence tends to lead to not writing down enough; neither is ideal. Among first-year undergraduates probably the more common tendency is to write down too much information, but whichever your tendency, the following suggestions will help:

- remember that a lot of notes doesn't necessarily equal good notes;
- look and listen for the important points—these are often the structural parts;
- use keywords to represent points or ideas concisely;
- add brief details of any examples or evidence that support a point.

Use abbreviations

Using abbreviations can be a real time saver. Use standard abbreviations, subject-specific abbreviations, and your own abbreviations for common words. The important thing is to be consistent and to ensure that your notes are still comprehensible. Don't use so many abbreviations that your notes turn into a shorthand transcript—these can be very difficult to decipher when your memory of the lecture has faded.

Use space

It can be tempting to try to cram as much information onto a page (or screen) as possible, but this will create difficulties for you both during and after the lecture. Notes that are densely packed are difficult to review and difficult to make additions to at a later stage. It is, therefore, helpful to use space in your lecture notes to make them easier to review and easier to supplement with additional material. To create space make sure you put each point on a new line (this also helps represent structure) and leave gaps for additions or corrections, especially if you think you may have missed a point or don't understand something.

Use colour and image

We have already identified the value of highlighting important points by using structure in the section on 'Use structure in your notes', but this can be further enhanced by using colour and image, too, as shown in Figure 3.3. It can be useful to highlight key points in colour, and to use images or diagrams as a quick way of summarizing a concept or idea. Sometimes such images or diagrams will be used by the lecturer, in which case, if it is useful, copy it down, although you are very likely to be provided with the slides to download, so you do not need to be able to draw well (see also the points in 'Use slides and handouts effectively')! On other occasions, you will think of ways to visually represent something that is only communicated verbally; this can also be a very helpful thing to do, but make sure you record sufficient information to be able to understand the concept or idea at a later stage.

Use slides and handouts effectively

Slides and handouts are a very valuable source of information. Many lectures won't have handouts, but most will have slides. Assuming that copies of the slides are available to you (either before or after the lecture) then the principles of using them effectively are similar.

- If you are using paper copies, put your name on the handout—everyone's will be identical (initially at least), and you will want to know which one is yours.
- Add the date and lecture title (if there isn't one on the slides or handout already) so you know which set of lecture notes it belongs to later.
- When making notes, think about what is contained in the handouts or slides, and don't copy down in your notes things that are already there. This will enable you to listen more actively and so focus on understanding.
- Make notes on the handouts or slides themselves, whether they are paper or digital—highlight important points, annotate, add comments, and write down any questions you have.

Figure 3.3 Using colour and image in notes

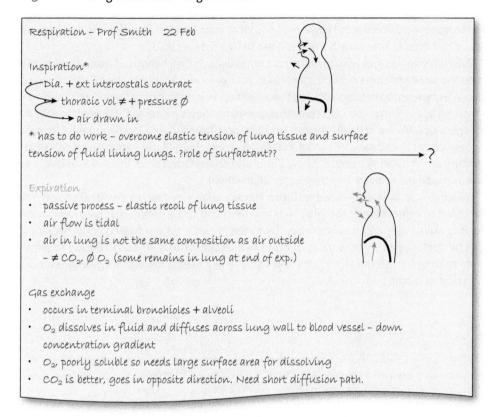

Respiration – Prof Smith   22 Feb

Inspiration*
— Dia. + ext intercostals contract
  → thoracic vol ≠ + pressure Ø
      → air drawn in
* has to do work – overcome elastic tension of lung tissue and surface
tension of fluid lining lungs. ?role of surfactant??   ————————→ ?

Expiration
• passive process – elastic recoil of lung tissue
• air flow is tidal
• air in lung is not the same composition as air outside
  – ≠ CO_2, Ø O_2 (some remains in lung at end of exp.)

Gas exchange
• occurs in terminal bronchioles + alveoli
• O_2 dissolves in fluid and diffuses across lung wall to blood vessel – down
  concentration gradient
• O_2 poorly soluble so needs large surface area for dissolving
• CO_2 is better, goes in opposite direction. Need short diffusion path.

- Some lecturers will provide 'gapped' handouts or slides where you need to fill in the gaps with key information. Follow the lecture so that you fill in the correct information in the gaps.
- Don't fall into the trap of thinking that because you've got a handout or slides that you don't need to listen much—listening actively during the lecture and making appropriate notes (on your handouts and in your own notes) will save you a lot of time later and also make it easier to understand the material.
- File your handouts and notes in an organized fashion (which brings us neatly onto the penultimate point in this section).

Organize your notes

It is important that you organize your notes effectively (whether paper or digital), as this will make it much easier for you when you follow up the lecture at a later stage, or use the notes for background reading for an essay or when revising for exams. If you are making notes on paper, the simplest, and in our experience the most effective, method is to use A4 paper for taking notes and to store your notes in a ring binder, together with the relevant handouts. To make ordering your notes easier, begin notes for each lecture on a new piece of paper and give them a clear heading of the lecture title, date, and name of the lecturer. Also add page numbers so you can order the pages easily. To store your notes and handouts in a ring binder you might need to hole-punch them—do this before they get lost!

Alternatively, you may wish to make your notes digitally on a laptop or tablet. If you do this then the same principles regarding structure, your own words, abbreviations, and so on still apply. Using a laptop or tablet also has the advantage that notes can be tagged, retrieved, and searched in a way that you can't do with paper-based notes (check out Evernote, Notability, Google Docs, Microsoft OneNote, and GoodReader). When using a laptop or tablet to make notes, you may need to work harder initially to ensure you still use space, structure, colour, and image (to avoid your notes being linear and characterless), but this will improve with practice (and the software is getting better all the time).

Whether you use paper-based or digital notes will be very much a personal choice. You will need to experiment to see what suits you best, and find a system that balances out the various advantages and disadvantages. Try to decide on a system fairly quickly, because it can be confusing to try to work from notes in multiple formats. If you do decide to switch note-making formats then it's a good idea to do so at a natural break point, like the end of a semester or module.

Try this—Comparing lecture notes

After a lecture, compare your notes with other people. Remember, we said earlier that the ability to make good notes is a skill that develops with practice; it is also a skill that develops as you see how other people do it. Comparing notes can be a helpful exercise because it can help you to:

- see how other people take notes and pick up good techniques;
- identify and fill in any gaps that you might have in your notes;
- discuss the content of the lecture and clarify your understanding;
- identify how the lecture relates to the rest of the module

3.4.3 Know the common problems and how to address them

We have addressed the ways to help you to make appropriate notes, but there will always be times when, for a variety of reasons, concentration dips and you begin to get left behind in the lecture. The trick here is not to despair and give up!

Failing concentration

You are much less likely to find your concentration straying when you use an active approach to note-making. By using your own words, and by using space, colour, and image, note-making will become a busy, but interesting activity. If you do miss some points because your attention strays, then just leave a space in your notes and check it out with the lecturer or another student later or, if lecture capture is available, review the recording of the lecture.

Being left behind

You may find that information is being delivered too fast for you to write down. If points pass you by, then again, leave a space and compare your notes with another student's or check the recording. Remember that doing some background reading for the lecture will help you to keep up as the information will not be entirely unfamiliar to you. Sometimes you can get lost because you don't understand the material that is being delivered. This may be the case for the occasional point or even for a large section of the lecture. Rather than giving up on the lecture, write a series of questions that you can try to follow up later.

3.5 Follow-up after lectures

At school or college, lessons usually cover everything a student needs to know about their syllabus, but at undergraduate level it is important to note that lectures don't tell you everything you need to know and so need to be supplemented with additional reading. It is vitally important that you are aware of this because, when you are revising for exams, if you only revise the lecture material your knowledge of a subject will be inadequate. We will deal with this more in Chapter 13, 'Getting the most out of revision'.

3.5.1 Know what kind of follow-up is required

In terms of the information covered, lectures are commonly used either to offer a broad overview of a subject (in which case your job is to fill in the detail) or to deliver detailed information on a subject (in which case your job is to fill in the background). There are, therefore, two issues to consider when identifying what sort of follow-up a lecture requires:

- does the subject require more depth (because the lecture gave an overview);
- does the subject require more breadth (because the lecture gave detail)?

Lecturers will often provide a reading list of books and research articles to help you with the follow-up. The extent to which you use these is up to you. For some articles, you may only need to read the abstract or the discussion, whereas in other cases, in order to understand the material, you may well need to read the whole paper. This will be particularly true later in your course. Do engage with this reading as it will help you provide additional detail or case-study information, etc., to assist your understanding of the lectures. You can also use the notes made from the reading (see Chapter 4, 'Working with different information sources') to supplement your lecture notes.

3.5.2 Ask questions

Don't be afraid to ask a lecturer for clarification, either during the lecture or afterwards. This can take a bit of courage, especially putting your hand up and asking a question in a room full of people, but it is highly unlikely that you are the only person in the room thinking that particular question. Also, lecturers will often welcome such questions, as they provide a bit of stimulus for them, as well as indicating that people in the room are paying attention. Additionally, you may be able to use tutorials to clarify or discuss material from the lectures.

3.5.3 Review your notes

Review your notes as soon as possible after a lecture, and make the most of your review by:

- highlighting points which seem particularly important or central;
- adding any details that you can remember from the lecture;
- showing links between points;
- correcting any mistakes;
- adding questions to highlight areas you don't understand or on which you need further information;
- supplementing your notes with material from the reading lists.

 Chapter summary

Making the most of lectures is an important skill for undergraduates. There is much you can do to improve your ability to listen and make notes; this includes things you can do before, during, and after lectures. The things you can do include:

- *Being prepared*: get to know the course structure and do pre-lecture reading.
- *Being organized*: have a system for storing notes (either paper-based or on a laptop/tablet), and take a selection of pens and pencils, or well-charged laptop/tablet to the lecture.
- *Listen for structure*: watch out for signposts that will help you follow the lecture.
- *Being brief*: try using key words and phrases as much as possible, so the emphasis is on listening and understanding.
- *Let note-making be an active process*: summarize in your own words, make good use of space, colour, symbols, and images.
- *Leave gaps*: if you miss a point, don't get further behind by panicking about it, just leave a space and check it out with another student after the lecture.
- *Actively review your notes after the lecture*: make additions and corrections as you read.

4 Working with different information sources

 Introduction

In previous generations, students were said to go to university 'to read for a degree', the implication being that most of their working time as students would be spent reading up on the subject they were studying. Even in subjects like the biosciences, which have a strong emphasis on practical work, reading is still a very important, if often underrated, part of studying.

What is the purpose of reading? This is an important question because there are several reasons why you will need to read during the course of your degree, and the approach that you take will depend on what your goal for the reading is. Some key reasons for reading include the following:

- background reading to find out more about the subject in general;
- following up on lectures, or other teaching activities, in order to help you understand the subject matter you are being taught;
- following up on lectures to develop a deeper and broader appreciation of the subject matter;
- researching topics as part of the preparation for a coursework assignment, such as writing an essay or delivering a presentation.

These activities require different approaches to reading because there are very different goals associated with them. For example, the sources that you use and the strategies employed for finding the information will differ, depending both on the goal for your activity and the level you have reached in your course.

This chapter explores how to get the most from texts, journals, and the internet, and how to evaluate the academic value of a variety of sources. To do this, we will review:

- the types of publication available and how you might use them;
- using internet sources;
- referencing and bibliographies;
- different reading goals and how you approach them;
- note-making strategies.

We will end the chapter by setting out an approach to critical review of a research paper, which is an exercise that you will almost certainly have to undertake at some point during your studies.

4.1 Different types of printed publication

There is a wide range of types of publication available for use at different levels of study. Each of these has different characteristics and will probably be used in different ways. In this section, we will consider the main forms of publication, some of their key characteristics as study aids, and suggestions as to their use. Further consideration of the use of publications and internet sources are given in the sections on note-making and reading goals.

A fairly simple way of categorizing the types of publication you might want to use for your studies is as follows:

- textbook;
- specialist topic book/monograph;
- general science journal;
- specialist review article;
- research paper.

Let us consider each of these in turn.

4.1.1 Textbooks

The textbook is probably the form of publication with which you will be most familiar from your studies at school or college. Textbooks still form a very important resource for students at university, predominantly in the first and second years of study. While many of your textbooks will have been broad-based (e.g. a textbook covering all of biology), the books you will be using, even in your first year of study, will be more specialized, focusing on a specific sub-topic of biology such as biochemistry, genetics, or physiology. The principle is the same, however—textbooks offer a broad-based foundation to the study of a subject.

Buying textbooks

Even within the individual topics, such as biochemistry or physiology, if you go to a bookshop or search online you will find a bewildering array of books available to buy. It is important to be cautious when purchasing books, otherwise you can find that you have spent a lot of money very quickly, sometimes on buying books that will not be of great use to you. You have clearly taken the first step in the right direction by buying this book, now you can proceed with caution!

- Most universities publish lists of recommended books for specific modules or courses, so start off by consulting the lists for the modules you are taking.
- *Be selective*: for example, is it worth buying a book for £40 or £50 for a subject area that you are taking as a single module in your first year, but which you do not intend to continue with later on in your course? If the university library has a number of copies available, you can probably manage without buying one for yourself.
- Having decided you do want to buy a book, you have to choose which one. It may be that your reading list only suggests a single core text, which makes life easy. However,

there may be a choice of recommended texts, perhaps covering the material to different levels and at widely differing prices. Again, the library is a useful starting point. Have a look at the relevant texts available in the library catalogue; you will probably find that they are different in terms of their layout, the way diagrams are used, the level of detail, etc., so you may find it helpful to test-read parts of the book to find which style suits you best. In the library, you will find that some of these books will be available as a hard copy, some as ebooks, some as both. You may also find that they differ in terms of the additional features they offer—for example, is there an associated website with more information or access to self-evaluation tests?

- You may consider buying a second-hand copy of the book. Major on line sellers are likely to have stocks of second-hand books for sale, especially if they are standard textbooks. Your university or Students' Union may also have a second-hand book shop or online alternative. The books should be cheaper than buying brand new so you could save money, but you need to be cautious. For example, make sure that it is the current edition of the book that is on sale. If the book has an associated website, make sure that you can still access it—many sites have access controls that make them unique to a single user and cannot be transferred if the book is sold on. Also, do check the shipping costs!

Using textbooks

Textbooks commonly provide a broad overview of a topic and can have several uses as study tools. Think about the reading goals that you have and that will guide you in the way to use your textbook. Some possible examples could be:

Goal 1

I need to broaden my knowledge and understanding of a specific topic, for example, the function of the kidney.

Action 1

In a general physiology textbook read through the chapter(s) on the kidney, checking on any learning outcomes identified in the text. Make abbreviated notes to aid your learning (see Section 4.3, 'Note-making strategies').

Goal 2

I have just had a lecture on the physiology of the kidney, which focused on the loop of Henlé and I don't understand it!

Action 2

Read through your lecture notes and any accompanying handouts. Look up the loop of Henlé in the index of your physiology textbook and read through the relevant section slowly. Then read it again, making notes about the organization of the loop and the way it functions. It might help to go to the library and read about it in a couple of different textbooks. You should also ask the lecturer for help, but it is a good idea to try to think it through for yourself first.

Goal 3

I have just been given an essay to write on the regulation of ion balance by the kidney. Where do I begin?

Action 3

Read through any relevant lecture notes and handouts first. Then read the sections in your physiology textbook that deal with ion balance, making detailed notes about each of the key ions in turn (e.g. Na^+, K^+, Cl^-, etc.). Make sure you put the notes in your own words and include the source reference (see Section 4.5, 'Citations and references'). Having got the general picture, you may then need to refer to more detailed sources, such as a specialist topic book, review articles, or research papers.

As with any source of information, textbooks do have limitations:

- Because most textbooks offer a broad coverage of the subject (e.g. physiology or ecology, etc.) they may well only provide a rather superficial overview of a topic. This is important because it may be just what you need for giving you the background to the topic, or for helping you understand lecture materials in your first and second years. They almost certainly will not, however, provide enough detail for the later parts of your course or for researching material for coursework.

- Textbooks are often written by a small team of authors, who will not be specialists in all of the topics covered in the textbook. As a result, the quality of the coverage of the different topics may vary. For some texts, the book is prepared by editors, with subject specialists writing the individual sections. These are more likely to have a more even quality of coverage of the topics.

- Textbooks do go out of date, and this is likely to happen faster than for your school or college textbooks. It can take several years for a textbook to be written and published, and then there may be a few years more before the book is revised and a further edition published. Although this may not cause problems when used for basic reading (since the key principles of a topic are unlikely to change significantly), if you require more specialist information, then it is more important that the book you are using is up to date. So, when you are scanning your library for textbooks, make sure that the books you select are not more than two or three years old.

4.1.2 Specialist topic books/monographs

As their name suggests, these are books that focus on a specific topic within the subject. These usually take three forms:

1. An advanced textbook dealing specifically with a fairly narrow topic, such as kidney function—this will clearly be more detailed in approach than the broader textbook on physiology, but will be written in a similar style. It will almost certainly have been written by a specialist, or group of specialists, within the field.

2. A monograph, which usually takes the form of a long article or a short book on a specialist topic—this will typically be narrower in focus and at a more specialist level than the advanced textbook.

3. At the research level, books based on the papers presented at a specifically themed conference—as with journal research papers, the individual papers may be specialized,

being written for a readership of research scientists within the field. However, the grouping together of a whole series of papers on a specific theme may well provide you with a valuable starting point for your research for an essay topic. These books frequently contain useful overviews of groups of papers in review form.

As with using textbooks, you need to make sure that the information you are looking at is up to date; indeed, the more specialized the book or research paper, the faster it is likely to go out of date. All of these, however, make valuable resources for advanced reading *around* a topic to increase your depth of knowledge, and they can also make very useful starting points in research for essays or similar assignments.

4.1.3 General science journals

There are several general science journals, such as *New Scientist* and *Scientific American*, which publish review-type articles aimed at the general interest reader with a scientific background. These types of journal can be very useful in providing overviews of a specific topic and also for keeping you up to date with wider developments in science as a whole.

4.1.4 Specialist review articles

Many research journals include review articles as well as the papers describing the results of novel research. There are also numerous journals dedicated to reviews in the biosciences, such as the *'Annual Reviews in . . .'* series or the *'Trends in . . .'* series.

Recent review articles are a gold mine for students seeking information about a specific topic. These articles are typically written by a well-known specialist in the field who brings together, in a single article, a synthesis of a large number of individual research papers. The approach of these articles is usually designed to enable someone with some knowledge of the field, but not necessarily at a very detailed level, to derive a picture of the current ideas and developments within that field, without having to read all the individual research papers.

Review articles are very useful for providing more detailed information about a topic, for example, for supplementing lecture material, particularly at second- and third-year level. They also often make very useful starting points in research for coursework assignments, both because of the overview they provide and because of the references made to recent research work in the field. Your lecturers will almost certainly include some review articles in the reading lists for your courses, particularly in the later years of your degree programme.

4.1.5 Research papers

Research papers published in academic journals are the dynamic bedrock on which current knowledge is based. These papers are written by the research scientists describing their latest discoveries or developments within their field of research. In any honours degree programme, there will be an expectation that you will:

- read recent research papers in order to supplement the lecture material;
- use them in order to prepare your coursework assignments;

- use the information you have acquired from reading research papers to support the arguments you present in examination essays (this will be particularly true for finals examinations);
- critically evaluate research articles as part of your coursework (see Section 4.4, 'Critically evaluating a research paper').

As with the review articles, your lecturers will certainly include recommended research papers (some of which they have probably written themselves!) in the reading lists they give you to supplement the material covered in the lectures. The fact that someone who is an expert in the field has recommended a given paper suggests that the paper has merit as a piece of scientific writing and that the findings are likely to be reliable. In the same context, papers for publication in scientific journals are subjected to a peer review process. In this process, reviewers (usually two), who are also acknowledged experts in their field, are asked to read the paper and to comment on its scientific value, to rate its importance in the field, and suggest any amendments they think should be made. Although this is clearly not a foolproof process, and is certainly not an excuse for you to stop thinking critically when reading, it does give these papers a stamp of authority and reliability.

Research papers are the most specialized of all the standard sources of information and can therefore be quite intimidating and hard to tackle: for example, they will probably use terminologies and be based on ideas which you do not understand very well. Research papers are usually in a highly stylized format, presenting information in a way to which you are probably not accustomed. Finally, it must also be admitted that they don't always make the most interesting bedtime reading! So, if you are going to invest time and effort in reading research papers, you need to be very clear about your reading goals so that you get the most benefit from your input.

The structure of research papers

Most research papers are written in a standard format, comprising seven main elements (see also Chapter 8, 'Writing practical and project reports').

- *Title*: the title is a short descriptor of the paper; as such it will give you a good idea of whether the subject matter is relevant to your needs.
- *Abstract*: the abstract is usually about 200 words in length and provides a short summary of the paper. In particular, this will include the aims of the research, the key results, and a brief commentary about the conclusions. It may be that this is all you need to read in order to glean the key points, for example, to support a specific point you want to make in an essay.
- *Introduction*: the introduction is the section of the paper where the authors set out the background to the study. This is often very useful both for helping you to understand the basis of the topic and the reason why the researchers undertook the investigation they are describing. As such, it will normally include a useful summary of the preceding research and an analysis of the questions still to be answered. It usually culminates in setting out the aims of the study.
- *Methods*: the methods section describes the way in which the research was undertaken, including details of the experimental protocols and the procedures for analysing the

data. Unlike the type of report that you might write (see Chapter 8, 'Writing practical and project reports'), the protocols are often not described in full, but rather refer to previous papers in which they have been described, for example:

> The muscle preparation was prepared for electron microscopy according to the method of Saito & Jones (2000) . . .

This form of summary information can be fine for people working in the field who already have a good knowledge of the literature. For the student trying to understand the paper, however, this can be a nuisance because it means further searching out of papers in order to find the actual description of the method, particularly if, as in this case, the method is an old, but standard technique.

- *Results*: in the results section, the authors will describe the results that they have obtained from their research and will also display them in different formats, for example, as graphs, tables, or other illustrations. Where quantitative data are being presented, this will also be analysed using statistical techniques. Note that this section is descriptive and does not normally include any discussion of the results or reference to the findings of other researchers.

- *Discussion*: this section is where the authors interpret the findings of their research, both in terms of the specific experiments they have undertaken and also in the context of the current hypotheses in the field, as developed by other researchers. This section should link back to the introduction, where the aims of the project were set out. Again, this section makes very useful reading as it links what has been reported in the paper to the existing research literature.

- *References*: this comprises a list of all the research papers, etc., which were referred to in the text of the paper. The reference list is often very useful in providing a source of further reading to help you fill in the background of your topic.

Reading a research paper

From the reading list supplied by your lecturer and your own searching of databases, you will probably have collected a frighteningly long list of research articles on a topic. Unfortunately (and contrary to some popular opinion), the act of downloading a paper or getting a copy from the library does not also transfer the contents of the paper to your brain. You actually have to read the paper in order to extract the information from it. You also know that there is no way that you will have enough time to read all these papers.

So what should you do? This is where you need to be selective in your searching (see Section 4.2, 'Using the internet') and think about your reading goals to decide what you really want to get from your reading. For example, is your goal to get some specific, factual information to add to an essay? Or is it to get a detailed understanding of the topic as the background to a final-year research project? It may be that just reading the abstract will suffice, or you may need to read the whole paper. These different approaches can be evaluated by careful consideration of your reading goals.

Goal 1

I am reading papers to find specific information for an essay or practical report.

Action 1

In this case, you are looking for specific information and do not necessarily need to read the whole paper. Check that the title appears relevant and then read the abstract—this should summarize the key points of the paper and may, by itself, be sufficient for what you need, particularly if the main use of the reference is to confirm a minor point in the essay. However, if the information you require is more detailed, then you will need to go deeper into the paper.

Goal 2

I am reading papers to support the lecture material on a module.

Action 2

For this type of reading, you may often begin from a reading list that you have been given along with your lectures. Therefore, you can start by selecting the papers from that list that appear directly relevant to the topic you are reading up on. As with Goal 1, you may be able to be quite selective in the extent of your reading, particularly if there is a specific point that you want to know more about. Alternatively, you may also wish to read the papers fully in order to appreciate the wider aspects of the topic. Try to make sure that, before you start reading, you are clear in your own mind why you are reading that particular paper and what you want to get from it.

Goal 3

I am preparing the critical literature review for my final-year project.

Action 3

For this type of exercise you will probably need to read the paper carefully, especially if you are comparing the hypotheses put forward by different authors. As with all your reading, when you get to a section relevant to your reading goal you should make careful, referenced notes about what you have learnt from each paper. This is particularly true when engaged in a long-term piece of work, such as your final-year project, because you may well need to be able to go back to the sources to check, or add to, what you have written.

So, how do you find this information if it is not taken from a reading list? The most common approach is to use internet-based searches, which we describe in Section 4.2.

4.2 Using the internet

Clearly, the internet contains a vast and bewildering wealth of information. Some of that information will be accurate, up to date, authoritative, and really useful. A lot of what you may unearth, however, is none of those things. So, while online searches can be very productive, you need to have a clear strategy for directing your search, for filtering the material that is available, and for being able to evaluate that information.

Let us say that you have been set an assignment on the mechanisms underlying muscle fatigue and you need to find out some recent information on the topic. If you need to start by finding out basic information, then it is probably best to start with lecture notes and hand-outs, progressing on to textbooks and reading lists as described previously. When you have exhausted these, you need to undertake your own specific searches.

There are two main starting points—you can look specifically for articles in academic journals, or you can use a search engine to explore the whole of the web. In this exercise, we will explore both approaches for comparison.

4.2.1 Academic journals

When you are searching for specialist scientific information, then academic journals should be one of your main sources of information. Almost all universities and colleges subscribe to electronic journals and also to databases that search the academic literature. Through your university library's catalogue, you should be able to find what journals you will have direct access to via an online link. The library catalogue should also provide links for you to use databases for your search. Examples of such databases that are commonly employed by bioscientists are Web of Science, Ovid, PubMed, and Medline. Different databases cover different ranges of journals, and so they vary in size and in terms of the number of journals included. Some, such as Web of Science, are relatively broad; others, such as Medline, are more specific, only including journals likely to be of relevance to medical research. You will need to select which database you use and should consider seeking advice from a university librarian, who will have a good knowledge of the databases your university subscribes to and the range of topics that they cover.

These databases allow very sophisticated search strategies using a range of criteria. The most common strategy is likely to be based on a set of keywords that you think are core to the topic. If you have been given the name of a specific researcher, you could also search using the author's name. The advanced techniques of searching are beyond the scope of this book, and you should consult with an adviser from your library, but we will look at some simple strategies for searching, which will probably be adequate for much of what you want to do.

You need to find out about the mechanisms of muscle fatigue so you could start with a crude search in Web of Science and simply enter the words 'muscle fatigue'. This comes up with more than 20,000 research papers. Similarly, if the same key words are entered into PubMed, more than 18,000 articles are listed. In both cases, you probably don't have time to scan the titles, let alone read the articles! So how can you refine the search strategy to pull out the articles that will be of value? There are several basic options here:

- choose specific types of article;
- select only more recent publications;
- refine your search terms.

See Figure 4.1 for an example.

Choose the type of article

Most search engines allow you to specify what type of article you would like to select. You can do this either by specifying what types you want to *include* or what types you want to *exclude*.

Figure 4.1 Example of page 1 of the results from PubMed search, searching on the term 'muscle fatigue' for articles within the past five years.

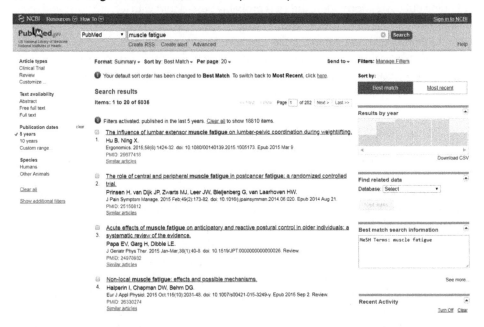

Copyright: Ovid Technologies, New York, NY.

For example, you can specify that you only want review articles, or you can exclude meeting abstracts (because they are likely to be short and very narrow in focus). If we only specify review articles, then PubMed comes up with about 2000 articles. This is still too many and there is a risk that you will miss out on some useful research papers that have not been covered by a review article. Furthermore, not all of these will be directly relevant, because of the breadth of the search terms used, so we need to refine our selections.

Select only more recent publications

Here, you can choose what years' publications you want to consider. This allows you, for example, to narrow down to a single year if you wish. A rule of thumb for your assignment might be to narrow your search down to the last five years as a starting point. If we select only the last five years, then the number of articles from PubMed drops from more than 18,000 to 4700. That's still too many, but it is a considerable improvement. There is a risk, however, that you may exclude some older, but very useful papers.

Refine your search terms

'Muscle fatigue' is a very broad term to use for searching, so you could consider refining your strategy. If you enter the term 'muscle fatigue' into the search box, then the database will search for all articles that include the terms 'muscle' and 'fatigue' but this does not mean that the words will necessarily appear together as a specific phrase. If you want to search specifically for 'muscle fatigue' then, in Web of Science, you enclose the term in quotation marks:

'muscle fatigue'

This will now only select those papers where the two words appear together. Doing that reduces the number of articles selected from 20,000 to 6700.

The title of the assignment actually referred to the 'mechanisms of muscle fatigue'. So a further refinement would be to include the word 'mechanisms', so your search term becomes:

'mechanisms of muscle fatigue'

This now reduces the number of articles to 20, five of which were published in the last five years and all of which are directly relevant. These will make a good starting point for the research.

This strategy worked fairly easily, but you may find that not all searches are as simple and that you need to be still more sophisticated.

A further refinement is to use what are called 'Boolean Operators'. The most common of these are AND, NOT, and OR. These allow you to specify combinations of search terms. For example, if you specifically want to find out about muscle fatigue research that has been done on humans you could enter the following string:

'muscle fatigue' AND human

This will now return a list of articles where the words muscle fatigue occur together and the word human also occurs. However, if you wanted to find out about any work that had been carried out on animals other than humans, you would enter the search string:

'muscle fatigue' NOT human

Through these types of search operators, you can create very specific criteria for what articles you want.

It may also be the case that you want to broaden your search strategy to include related terms, particularly if your initial search has only returned a small number of articles. You could think of all the variations of the key term and type each one into your search phrase, which is not a very efficient approach, or you could use a truncation symbol such as *. If you enter the search string:

Musc* fatigue

this would return all the terms with 'musc' as the base, such as muscle and muscular. This can be very useful, but again there is a risk of getting too many responses, not all of which will be relevant. For example, if rather than 'musc*', you had shortened your term to 'mus*' this would return instances of muscle and muscular, but also all the other words that have 'mus' as the base, for example mush, must, musk, etc.

Another way of broadening the search terms to include variations on a word is to use a wildcard symbol such as ?, where the ? replaces one or more letters. For example, 'wom?n' would return both 'woman' and 'women'.

Each time you undertake a search, you should look at the number of articles that the search has returned and scan the first few titles to check that they appear to be relevant. With a broad search you are likely to find a large number of articles, many of which are not directly relevant. As you refine your search, for example, by selecting specific years or by using more specific search terms, then the number of articles should progressively reduce and should become more relevant, but there is a compromise as you may also start to exclude some articles that would be useful.

Coming further up to date

In your reading list, or from your searches, you may have an article that is very interesting, but is four or five years old. So you want to find something more recent that is directly related. If you have obtained the details of the article from a search engine you can use that article as a starting point and search using citations. In other words, the database search looks for more recent articles that cite the one you have obtained. This enables you to bring your research up to date, following the specific theme of your original article.

Try this—Practising search terms

Think of a topic that you have just covered in your course and set yourself this test:

1. Find five review articles on the topic that were published in the last three years.
2. Then find five articles published within the last three years that address a specific aspect of the topic.
3. In doing this exercise, experiment with the search terms using some of the processes covered in the preceding section.

Try the exercise in two different databases and see if you end up with the same articles!

4.2.2 Web search engines

The major web search engines, such as Google, Yahoo!, Bing, etc., are familiar to most people and they can be used to scan literally billions of pages of web-based information. They are very effective at producing long lists of websites related to your inquiry, but this means you have to be very selective in terms both of finding material that is directly relevant and in evaluating the reliability of that material. Don't forget that anyone can host a site and put information onto the web, and there is no guarantee that the information has been checked for accuracy.

Researching for our essay on the mechanisms of muscle fatigue using search engines, we can adopt the same procedures as when using the academic journal databases. If we start simply by entering the key words:

---

mechanisms muscle fatigue

Google returns 78,000,000 results, Yahoo! returns 274,000, and Bing returns 278,000, and all three engines ranked Wikipedia (see section on 'How reliable is a website?') near the top of the list. The ranking of the list, however, does not necessarily indicate the direct relevance of the information to be found on the site (see Box 4.1). You will also see a lot of different adverts on the pages, for example, for dietary supplements to help improve your muscle strength!

Although the high number of sites means that there is only time to check the first few, a scientific topic such as the mechanisms of muscle fatigue tends to come up with a significant number of research papers from academic journals and links to other academic sites, such as universities in the first listings, so these are likely to be reliable sources for your essay.

If the title is on a more topical subject, the results are very different. For example, consider the 'effects of smoking' as a title. Using the keywords 'effects' and 'smoking', Google returned 283,000,000 results!

This means that, although the first pages in the listing will be popular, they are not necessarily the most relevant pages and you may need to scan quite a way through the listing to pick out the most relevant sites. Again, therefore, you may need to consider refining your search terms; for example using your title phrase in quotation marks: 'effects of smoking' reduces the number of hits to 15,400,000.

These links will be to a whole variety of sites, such as commercial sites offering products to help you stop smoking, government advisory sites, pressure groups (from various perspectives), media sites, such as the BBC, charities and self-help organizations, medical advisory groups such as NHS Direct, and many more. Therefore, you will have to exercise judgement in terms of the types of site you look at and you need to determine their reliability (see the section on 'How reliable is a website?').

It may be that, for your particular task, you need a range of sources of information, not just research articles. You can therefore use the broad search engines and perhaps refine your

BOX 4.1 Web search engines: a very brief introduction

When we are looking for information about muscle fatigue and enter those words into the search box of a search engine, the search engine will return sites that contain the words 'muscle fatigue', but the words will not necessarily be together, or in that order, unless you have used an advanced search strategy. The order of the listing of the sites is also determined by a ranking process. This process ranks pages on a number of criteria, one of the most important of which is the number of 'backlinks'. These are the links, for example, hyperlinks in text, feeding to this page from other web sites, so the more links there are to a given web page and the higher the ranking of each of those sites, then the higher that web page will appear on the search engine's listing. The use of specific keywords also contributes to the ranking algorithm, as does optimization for mobile devices.

search using the advanced search techniques available. As with the academic databases re-ferred to earlier, these advanced search techniques allow you to specify search criteria, such as by an exact phrase or without specific words.

If you want to search for academic articles only, then the search engines now include sub-sections that will do that. An example is Google Scholar—a search engine that searches peer-reviewed papers, books, and conference proceedings. Although the range of journals that can be accessed by these types of searches is not always as extensive as for the specifically aca-demic engines, there has been a rapid expansion, and search engines such as Google Scholar do offer a good alternative. As with the specific search engines such as Web of Science, with Google Scholar you can refine your search strategy by combining terms and being selective about the year of publication, etc. A further approach that can be productive is that you can search on the basis of citations; for example, if you have a research paper from a few years ago as a starting point, you can search for all the papers that have cited that paper in recent years. This enables you to track a specific theme up to the present day.

How reliable is a website?

As we have seen, there are many sites providing information on specific topics and some are more reliable than others. So how do you know what to use? From a purely academic stand-point, the scientific journals, with their peer-reviewed papers, offer the most reliable sources of information, but there are many other useful sites as well. The simplest questions to ask yourself about each site are who or what organization wrote the material, and how old is it?

You can quickly answer these questions by looking at the site address in the listing. Each site is identifiable by its universal resource locator (URL). Examples of useful sites would include:

- *University sites*: these can include course materials from the university's undergraduate programmes, which can be very useful as an additional resource, as well as providing links to the publications and research outputs of their academic departments. Sites for higher education institutes are identifiable from the URL, which will contain the code ac (academic) or edu (educational). Thus, universities in the UK all have an address that takes the form: www.[universityname].ac.uk. Those in the USA have the address www.[universityname].edu, and for Australia they are www.[universityname].edu.au.

- *Government sites*: these can provide information at a range of levels, particularly in terms of policy or public advice. Examples here would be the Food Standards Agency, the Environment Agency, or the National Health Service (NHS). Most government sites in the UK have the address form that begins with www.gov.uk followed by the department title, or www.[title].gov.uk; NHS sites similarly have the address form: www.nhs.uk/[name] or www.[name].nhs.uk.

- *Learned societies*: there is a learned society associated with almost every specific area within the biosciences; for example Society for General Microbiology, Physiological Society, British Neuroscience Association, Society for Experimental Biology, and many more. Most of these sites have addresses in the form: www.[name].org. Addresses ending in .org are used for organizations that are non-governmental and non-profit-making.

- *Professional bodies*: these are organizations that represent and maintain the professional standing of specific professions, such as the General Medical Council, the General Dental

Council, and the Royal Society of Biology. These sites, too, usually have addresses as: www.[name].org.

- *Media organizations*: many of these organizations, such as the BBC, *The Times*, *The Guardian*, and *The Daily Telegraph*, are very useful for providing information about topical science stories. Remember, however, that the story is usually written up by a journalist who may have a limited knowledge of the topic and is writing for the general public, rather than for someone looking for detailed, scientific information. Media organizations are also in the business of selling news stories and, therefore, they may focus on the more sensational aspects of a topic.

- *Commercial organizations*: many commercial organizations have very useful websites, but it is important to remember that they also want to sell a product. Going back to our search on the effects of smoking, there were numerous sites listed that were hosted by commercial organizations selling products to assist in stopping smoking. These carried quite a lot of useful educational information, for example, descriptions of the dangerous effects of smoking on the lung and cardiovascular system. However, they also carried a lot of commercial information setting out the benefits of their specific product. Although the information published by such sites has to comply with the requirements of the Advertising Standards Agency, they do not have to be objective in their approach. For example, while they are extolling the benefits of their own product, they are under no obligation to inform you that another company's product may actually be more effective.

- *Specific societies and organizations*: there is a huge variety of organizations and groups with websites, and the quality of these varies greatly. In the area of health, there are numerous self-help organizations that have sites that carry useful information, such as the Parkinson's Disease Society, the Alzheimer's Society, or the Motor Neuron Disease Association. In environmental biology, there are organizations such as the World Wildlife Fund and Greenpeace. These sites are typically aimed at a non-academic readership, but often carry a significant amount of background information. However, these organizations frequently have very focused agendas and, therefore, the information provided may be selective or interpreted in very specific ways, so when you are reading, you need to be aware of that focus and to maintain your critical approach.

Mention must also be made of one of the most used sources of information on the web, Wikipedia. Wikipedia is an online encyclopaedia, but it differs from all other encyclopaedias in that anyone can add or delete information from specific entries. This means that, whereas much of the information present may have been written by experts, and should be accurate and current (often more accurate and current than traditional encyclopaedias), an entry may also have been written by someone with very limited knowledge and may contain major inaccuracies. Unless you already know about the topic, you will not know whether the content is accurate or not. Wikipedia can be useful, for example, for reminding yourself quickly about a topic or as a starting point for some research, but it is not necessarily advisable to use it if you are learning about a topic from scratch or researching for an essay. Your lecturers are unlikely to be impressed if the reference list for your assignment contains articles from Wikipedia.

4.3 Note-making strategies

It's past midnight and you are finally in the process of writing your essay. It's due in for to-morrow morning, but that's okay because you have done all the research, you have your plan in front of you, and the writing is going well. Then you hit a problem—you just need to describe the key experiment that supported the argument you are putting forward. You know you read up on it, but you can't quite remember the details. It's somewhere in that pile of review articles, but you can't remember which one and you don't have any notes to jog your memory . . .

Situations like this are very frustrating, but they can easily be avoided if you are systematic in making notes when you read up on a topic. Many students only think of note-making in the context of lectures, but it is just as important in terms of your reading around the subject. We can divide note-making into two processes. The first is generic and comprises the things you should be doing irrespective of why you are making the notes. This includes making sure that the information you are using is properly referenced so that you avoid plagiarism (Chapter 5, 'Academic integrity and avoiding plagiarism'). The second process varies depending on what you want to use the notes for, for example developing understanding of a topic, revision purposes, or information for a coursework assignment.

In the following sections, we consider each of these in turn.

4.3.1 Generic note-making

To ensure that your notes are as useful as possible and that your note-making is done effectively, you must always record the source and make sure that your notes are notes.

Record the source

Always make sure that you record the source of the information. You never know when you may need to find it again, particularly when writing your essay in the early hours of the morning; the more details you give, the easier life will be later on (Figure 4.2). For example, just jotting down the title of a book is of limited help, particularly if it is several hundred pages long! Don't forget also that if you are note-making for an assignment, you will need the full reference for your reference list.

Notes are notes!

You should always aim to note the information in your own words—just copying a paragraph out, or photocopying it, is not note-making. The process of reading and then summarizing the information in your own words is a very useful exercise because:

- as an active process, it will help you retain the information in your memory;
- it also tests your understanding of the information;
- it makes it faster for you to retrieve the key information at a later date, for example when revising.

Figure 4.2 Reference details for notes from a book and a research paper. It's okay to use abbreviations in the details since you can go back to the full reference if you need it for your assignment.

Dimmock, N. et al (2016) Intro. To Modern Virology
(7th ed.)

Chap 26, p411
    Principles of vaccination....

Crowe, J. (2017). Principles of broad and potent
antiviral human antibodies: Insights for vaccine
design. Cell Host & Microbe 22(2) 193-206

    Structure of antigen-antibody complexes linked to
vaccine design...

Have a look at the extract from the sample text in Figure 4.3 and then the notes in Figure 4.4 taken from the sample text. Identify some of the techniques used to set out the notes. These include:

- use of highlighting to emphasize the topic;
- leaving space to keep the notes clear;

Figure 4.3 From Nordell, S. and Valone, T. (2015). *Animal Behaviour: Concepts, Methods, and Applications*. With permission of New York: Oxford University Press

Many bird populations are declining because of human exploitation and habitat change. Natural habitats are often converted for the purposes of urban developments or agriculture. After conversion, they may no longer be suitable for particular species or may contain novel competitors or predators. In response, conservation biologists are working to restore degraded places and even create new habitat areas. Once a population has left a location, how do you attract it back to the new or restored habitat?

One method increasingly used to restore bird populations takes advantage of conspecific attraction. For instance, populations of colonial nesting seabirds such as Atlantic puffins (Fratercula arctica) have been restored to former breeding grounds by placing models and dummy nests at former breeding sites to attract individuals (e.g. Kress & Nettleship 1988; Parker et al 2007). Song playback has also been used to attract species back to areas.

Figure 4.4 Example of notes taken from a piece of text

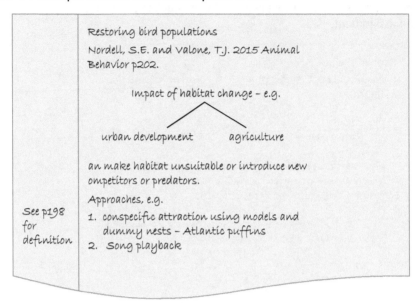

- leaving a wide margin, so you can add additional notes or explanatory comments later on;
- the source of the material (including page number) is given at the start of the notes;
- lines are used to indicate linkages of ideas and the division of the types of disorder into two groupings;
- the writing is in a very abbreviated style: there is no need to write in full sentences or include words like 'the', etc.

4.3.2 Focused note-making

As with your reading goals, the way in which you make notes will depend, to some extent, on the reason why you are making the notes. Some possible examples are:

- to aid with revision;
- as a tool to help with learning about a new topic;
- to aid understanding of a topic;
- as part of the research for a coursework assignment.

Assuming that you will be adopting the generic approaches described in Section 4.3.1, 'Generic note-making', how do these goals differ in terms of note-making? You could look at them in terms of your requirements. For example, the first two goals are likely to be fairly broad in coverage of a topic, whereas the last two are likely to be more focused on a specific aspect of a topic.

Note-making to aid with revision

You are preparing for an exam and have a large volume of notes from lectures, some background notes from reading textbooks, and perhaps some notes from research papers. It is a very useful exercise to bring these notes together so that they form a coherent theme that you can read through, rather than leave them as chunks of information dotted around (Figure 4.5). As well as making the notes easier to revise from, the process of actually undertaking such an

Figure 4.5 Construction of revision notes from lecture notes and notes from a text

> Muscle Organization   Dr Bond   Lecture 7   23/10/18
>
>   Motor units – functional grouping of m fibres
>   MU = 1 α motor neuron + group of mfs.
>   Size of MUs and contract. properties varies – 3 main groups – S, FR + FF
>   Postural muscles, e.g. soleus – mainly S
>   Muscles for jumping etc. –
>   gastroc. – mainly FF

> Motor units
> Human Physiology, Davies et al, p205
> S Units – slow, weak contractions, oxidative metabolism, don't fatigue
> FR Units – fast, intermed. contractions, ox/glycolytic metab, fatigue resistant
> FF Units – fast, strong contractions, glycolytic metab, fatigue quickly

> Muscle Organization   Revision notes
>
> Motor unit – functional grouping of muscle fibres
> Motor unit = α motor neuron + the group of muscle fibres it innervates
> MU size + contractile properties vary
> 3 main groups of MU: S – slow contracting, weak, oxidative metab.,
>   v. fatigue resistant
>   FR – fast contracting, stronger,
>   oxidative/glycolytic
>   Fatigue resistant
>   FF – fast contracting, strong, glycolytic,
>   fatigue quickly
> Postural muscles, e.g. soleus – mainly S
> Muscles for jumping etc. – gastrocnemius – mainly FF

exercise also helps you to learn. This topic is addressed in more detail in Chapter 13, 'Getting the most out of revision'.

Note-making as a tool to help with learning about a new topic

It may be that you have to research a topic from scratch, rather than using note-making as a supplement for your lecture notes or handouts. In this case, the best approach is to build up your notes in stages, rather like the revision notes. To start, you need to get an overview of the topic, for example, by reading and making notes from a general textbook (see Goal 1 under 'Using textbooks' in Section 4.1.1, 'Textbooks'). Then go into the topic more deeply using a review article or research papers to focus on specific aspects. Finally, produce a set of composite notes by combining the overview notes with the more detailed ones. By the time you have done that, you should have a useful set of notes, but should also have gained a good appreciation of the topic. Furthermore, when you come to produce the composite notes, it will also become clear to you if there are still areas that you do not understand.

Note-making to aid understanding of a topic

This task is a more focused note-making activity since you are likely to be concerned with a small aspect of a topic that you do not understand. In this case, your reading will be very goal-directed (Goal 2 under 'Using textbooks' in Section 4.1.1) and you only need to make short notes to supplement your current notes from your lectures. In this context, it is very helpful if you have large margins or space in your lecture notes (Chapter 3, 'Making the most of lectures') because then you may be able to insert the additional information within the original notes, rather than having to write it separately (or if you are making digital notes you will be able to slot in additions more easily). Amalgamating notes from different sources is a very useful exercise, as discussed earlier in this section, but it is not necessarily such a good use of your time if there are only very small sections of additional information that need to be included.

Note-making as part of the research for a coursework assignment

As part of the preparation for your assignment, you should have drawn up a list of specific topics that you need to address (see, e.g. Section 7.3.4, 'Make a plan' in Chapter 7, 'Writing essays'). Under each of those headings, you can now set out the key pieces of information from your reading, which will supplement other sources such as your lectures. As for understanding the topic, each point is likely to be quite brief and you will probably only need to record three pieces of information:

- the factual information itself;
- a summary of the experimental approach used to find the information;
- the full citation for the reference (see Section 4.4, 'Critically evaluating a research paper').

Try this—Checking your notes

Have a look at the most recent sets of notes that you have made for an assignment and check the following questions:

1. Do the notes summarize the key points clearly?
2. If I re-read the notes in a few weeks' time would I still understand them?
3. Have I noted the sources of information used so that I could find the specific points again?

4.4 Critically evaluating a research paper

Critical evaluation is a very useful skill to develop both for supporting your studies, and in your future life and work. If you are going to be successful as a bioscientist you must be able to evaluate your own work and that of others so you can identify the key questions that need to be asked and whether the answers given through research do actually answer the questions being posed. In everyday life, we are continually being presented with information and interpretations of that information—think of political debates, news articles in the media, or advertisements. All of these set out to justify a specific position based on evidence, but it is very important that you don't take this on trust; you need to look critically at the evidence, how it was collected and analysed, and whether the interpretations are actually justified. Do most dentists really recommend this toothpaste and what does 'most' mean? Is the job market actually improving or getting worse? Will this herbal remedy actually make you better?

As we discussed in Section 4.1.5, 'Research papers', in all honours degree programmes there will be an expectation that you will make use of recent research papers as an integral part of your studies. This will take the form of reading these papers to supplement the material you have covered in your taught sessions and to enable you to produce your coursework assignments.

You may have been given a list of papers to read, but you will also need to search the literature to find relevant papers to support your work. When you have found a possible set of papers through the use of search engines (Section 4.2, 'Using the internet'), you will need to select the papers on the basis of their relevance, but also their quality. To do the latter you will need to critically evaluate the paper. It is also very likely that you will be asked to critically evaluate a paper as a specific assignment task.

In many ways, critical evaluation of a paper is very similar to the process of writing a research report (Section 8.3, 'Writing your report'), because many of the questions you need to ask about the structure and individual sections are the same. The first step is to make sure that you read the paper through carefully and make notes as discussed in Section 4.3, 'Note-making strategies'.

Do bear in mind that, in common understanding, we tend to use the terms 'criticize' or 'critical' in negative contexts, implying that something is wrong. In this context, one of the definitions given in the Oxford Dictionary is really useful: being critical involves 'the objective analysis and evaluation of an issue in order to form a judgement'.

When analysing a paper, it is often helpful to read all through the paper first of all to obtain an overview, and then review each section in detail and attempt to answer the questions set out in Box 4.2. Make sure that you take notes as you go along: you may find it helps to fill out a Table template, such as the one provided in the website that accompanies this book: www.oup.com/uk/johnson3e. You don't need to give very detailed answers to each bullet point but, particularly if you identify errors or failings in the paper, you will need to give some evidence to back up your conclusions.

BOX 4.2 Steps to evaluating a research paper

(A template is available as part of the online resources at www.oup.com/uk/johnson3e)

1. For your critical analysis you should begin with questions about the reliance that might initially be placed on the paper:

   - When was it published: i.e. is the research likely to be current?
   - Where was it published: has it been subject to peer review? Is the journal highly rated in terms of its impact factor? You may need to go to the journal's website to answer these questions.
   - If it is a paper citing trial data, such as testing a pharmaceutical agent, have the authors made a declaration of having no vested interest in the outcome? For example, the authors should confirm whether or not the study was sponsored by the pharmaceutical company that manufactures the agent.

2. The next step is to take an overview of the paper, including the following points:

   - What is the research question that is being addressed?
   - What is the case made for the significance of the research: why did the author(s) consider it to be important?
   - What was the main experimental/observational approach, e.g. how were the data collected?
   - What were the main findings?
   - What were the key implications of the findings in terms of advancing current knowledge?

If we assume that the paper is set out in the standard format, then the next steps are to review each section in turn.

3. *The Introduction*

   - Does the introduction set out a clear background to the study in the context of current literature?
   - Is the question to be addressed clearly identified?
   - Do the authors explain why the research was done, i.e. why it was important to address this question?
   - Do the authors establish a research hypothesis that they aim to test?

4. *The Methods*

   - Do the methods give a full description of how the experiments/observations were undertaken and the data collected? Could someone repeat the experiments based on the information provided?
   - Are there citations to the literature identifying the sources of the protocols that were used?
   - If the protocols identified above were modified, is it clear how and why this was done?

(Continued)

    ● Have the authors commented on any limitations to the methods that were employed and how these were allowed for?

    ● Have the authors stated what statistical methods were used to analyse the data and why they were selected?

    ● If the research requires it, for example, for experimental procedures involving humans or other animals, have the authors obtained full ethical consent for the work?

5. *The Results*

    ● Are the results clearly and logically presented through the use of text, tables, and figures?

    ● Do the data presented in the tables and figures support the statements of findings made in the text?

    ● Do the statistical analyses support the statements of findings made in the text?

6. *The Discussion*

    ● Have the authors effectively interpreted their results in the context of the current literature?

    ● Do the findings, along with analysis of the context, provide sufficient evidence to enable the authors to accept or reject the initial hypothesis?

    ● Has the significance of the findings been clearly stated and explained?

    ● Are the authors justified in the claims they are making for the significance of the research and the conclusions they have drawn?

7. *References*

    ● Do all the references match with the in-text citations?

    ● Have the authors cited a relevant range of references, including current research?

Having read through the paper carefully and addressed these questions, go back to the title and the abstract.

8. *Title & Abstract*

    ● Does the title give a clear indication of the nature of the work?

    ● Does the abstract give an accurate summary of the work done?

    ● If read on its own, would the abstract allow the reader to understand the question(s) being asked, the core findings, and their significance?

    ● Are the summary statements of findings and their significance justified in the detailed descriptions in the paper?

Finally, you should draw all these points together in a summary statement of your conclusions regarding the quality of the paper.

Try this—Critically evaluate a research paper

Find one of the research papers that you used for a recent assignment and work through the list in Box 4.2 to see how you would rate the quality of the content and presentation.

    You might also like to have a look at some critical reviews of news articles, which pick up on biomedical stories and evaluate their reliability. *Behind the Headlines* is a resource published by the NHS that comments on the way in which biomedical science is reported in the media: https://www.nhs.uk/news/.

4.5 Citations and references

When you are writing or presenting any piece of work, it is very important that you acknowledge the sources of the information and ideas that you are presenting. If you present an idea or a piece of information without acknowledging a specific source, then you are effectively claiming ownership of that idea or information. Provided it is your idea then that's okay: for example, it could be your interpretation of the results obtained in a practical class or from your research project. The failure to acknowledge sources, however, is one of the most common reasons for being accused of plagiarism (see Chapter 5, 'Academic integrity and avoiding plagiarism').

In this section, we will look at a standard way of referencing, and discuss when and how you should do it.

4.5.1 Defining the terms

The process of acknowledging sources sometimes seems more complicated than it need be because of the different terms used to describe what you should do. We will work through some examples, but it is useful to explain some of the terms first.

Referencing is the name given to the process of acknowledging any material in the form of ideas, specific information, drawings or illustrations, experimental methods, computer programs, etc., that has come from someone else's work. Normally, this is work that has been published in some form, for example as a research paper, book, or film, but it may also be simply a theory that someone has told you.

Citations are the specific acknowledgements to other people made, for example, in the text of your writing. Normally, these will be the name of the person and the date when their material was published: see some of the examples in the short Section 4.5.4, 'How to draft citations and reference lists'.

References are the list of sources that you actually referred to in your work. Every citation in the text should appear in the reference list and every source listed in the references should refer to a citation in the text.

Bibliographies are lists of all the sources that you made use of during your research. A bibliography may, therefore, include sources such as general textbooks that you used for background reading, but from which you did not actually take specific material for your work. Some of the material listed in the bibliography may not, therefore, appear as a citation in the text. For most work at university level in the biosciences, particularly in the later years of your course, you will be expected to produce a reference list, rather than a bibliography.

4.5.2 Why should I cite other people's work?

The main reason, as indicated in the introduction to this section, is so that someone reading your work will know where the ideas you are presenting came from and will be able to identify which ideas or pieces of information that you are presenting are specifically your own. The references are also very helpful to the reader who needs to know more about the background to the topic. He or she can go to the articles you have cited to find out more detailed information. When you are preparing coursework, the list of references will show your tutor what materials you have used, and he or she may be able to give you feedback about the suitability of the materials and your search strategies.

4.5.3 Do I have to cite everything?

The general principle is that anything that you include in your work that is taken from someone else must be acknowledged. This is true just as much if you are quoting someone word for word as if you are taking their ideas and putting them in your own words. Having said that, there is a level of information that is now accepted knowledge and for which it is not necessary to provide a citation. For example, if you write the statements:

Chlorophyll is the main pigment involved in photosynthesis in green plants.
Bats are mammals that can fly.
Haemoglobin is the respiratory pigment involved in oxygen transport in humans.

These are accepted facts and there is no need to cite the source of the information. Indeed, the derivation of such information is often unclear anyway! To get a feel for how much you should cite, one of the best things you can do is think about it when you are doing your background reading—when you read a review article or a research paper, notice the level of information that the authors are taking as accepted and the levels at which they make specific citations. Have a look at these guidelines as a framework for knowing when to cite your sources.

You should always cite the source when:

- you are quoting word-for-word from a source, whether it is a textbook, research paper, lecture, or film. Any such quotation must also be written in quotation marks;
- you are copying, or taking elements from, a figure from a piece of published work;
- you are presenting information or ideas that are critical to the arguments you are making;
- you are presenting any ideas that may be controversial, for example, someone's theory about a specific process;
- you are presenting information based on specific experimental or observational evidence.

If in doubt, then acknowledge the source.

4.5.4 How to draft citations and reference lists

If you look through different academic journals, and particularly if you look at publications in different subject areas, you will see that the citations and reference lists are formulated in a variety of ways. There are several standard styles for referencing such as Harvard, Vancouver, Chicago, and so on. Although the styles are different for these, the basic information is always the same. In this context, Harvard is one of the most common formats for citations and reference lists in the biosciences, so this is the format we will employ in this text.

In the Harvard system, the citation in the text simply comprises the name(s) of the author(s) and the date. In the reference list you must also include the author's (or authors') initial(s), the title of the article and the journal or book it comes from, the volume number, and the page number. In the following examples we will run through the most common variations for citations and references. Note that, even within the Harvard system, there are differences in presentation. The following formats are based on the British Standard (BS ISO 690:210), but you should check with your lecturers whether your department recommends any specific styles. In the sections below, we cover the main principles of using the Harvard system, but detailed guides are available at:

- https://www2.le.ac.uk/library/help/referencing/author-date
- https://www2.le.ac.uk/offices/ld/resources/writing/harvard.

Referencing journal articles

For much of your study, articles published in scientific journals will represent the major source of information. Note that the following formats for referencing apply to all journal articles, even if you downloaded the article from the journal website.

Citation:

If there are up to three authors for a given publication, then all the authors' names are given:

Or:

Brie and Camembert (2017) reported that the moon is made of cheese.

Reference:

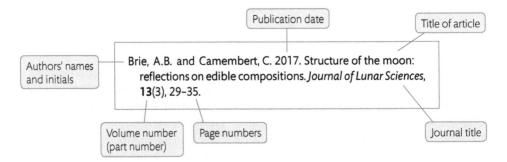

If there are four or more authors, only the first name is given in the citation followed by *et al.* (this comes from the Latin expression *et alii*, meaning *and others*), but all the authors should be listed in the reference.

Citation:

> The moon is made of cheese (Brie *et al.*, 2017).

Or:

> Brie *et al.* (2017) reported that the moon is made of cheese.

When giving a direct statement some authors prefer to use a phrase such as:

> Brie and co-workers (2017) reported that . . .

To cite two different papers by different authors in relation to the same statement:

> The moon is made of cheese (Brie and Camembert, 2017; Cheddar, 2018).

Or:

> Brie and Camembert (2017) and Cheddar (2018) reported that . . .

If you have more than two papers by different authors, listing the authors and dates in brackets is usually preferable to a long list of names in the body of the text. Note that the publications are normally listed in chronological order.

To cite two papers by the same author(s) in different years, list the papers in chronological order:

> The moon is made of cheese (Brie and Camembert, 2015, 2017).

To cite two papers by the same author(s) in the same year, the papers should be identified as 'a' or 'b' and this distinction added to the reference list.

Citation:

> The moon is made of cheese (Brie and Camembert, 2017a,b).

References:

> Brie, A.B. and Camembert, C. 2017a. Structure of the moon: reflections on edible compositions. *Journal of Lunar Sciences*, **13**(3), 29–35.
> Brie, A.B. and Camembert, C. 2017b. Variations in the density of cheese on the lunar surface. *Journal of Lunar Sciences*, **14**(1), 54–63.

The assumption so far is that these journal articles are articles that you have read, i.e. that these are *primary* sources. It may be that you wish to cite the work of a researcher, but have not read the material in the original article—for example, if you have not been able to obtain a copy of the article. In this case, you may have found that the information you need is quoted in a more recent paper by another author, and so you can refer to this more recent paper. This is clearly more indirect than taking your information directly from the original research paper, and it is referred to as a *secondary* source. The use of such secondary sources is not ideal, but sometimes it is unavoidable, particularly when you cannot access the original article. The format for such referencing is as follows (note that the reference is for the source from which you extracted the information, not the original source).

Citation:

> Cheddar had originally proposed that the composition of the moon was of blue stilton (Cheddar, 1994, as cited in Brie and Camembert, 2017).

Reference:

> Brie, A.B. and Camembert, C. 2017. Structure of the moon: reflections on edible compositions. *Journal of Lunar Sciences*, **13**(3), 29–35.

Referencing books

If you wish to cite material from a book, the citation in the text takes the same format as for the journal articles: the name(s) of the author(s) and the date of publication. The reference list format varies according to the type of book. The basic format is as follows:

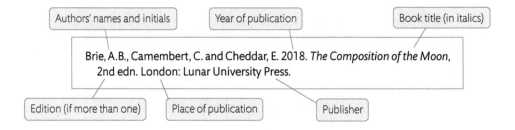

Many books are compositions of articles that have been edited by the main editors. In this case, the reference would be:

(eds) indicates that the named people edited the book, rather than actually wrote it.

Brie, A.B., Camembert, C. and Cheddar, E. (eds) 2018. *The Composition of the Moon*, 2nd edn. London: Lunar University Press.

In the case of such edited books, you may have only cited a specific chapter in the book that was written by another author. The format for such a reference is as follows:

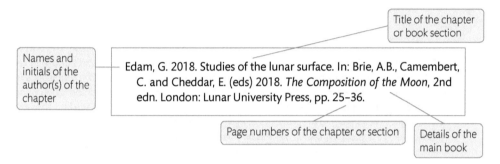

Title of the chapter or book section

Names and initials of the author(s) of the chapter

Edam, G. 2018. Studies of the lunar surface. In: Brie, A.B., Camembert, C. and Cheddar, E. (eds) 2018. *The Composition of the Moon*, 2nd edn. London: Lunar University Press, pp. 25–36.

Page numbers of the chapter or section

Details of the main book

Referencing websites

An increasing amount of information is available from websites and many such sites can be very valuable sources (though see Section 4.2, 'Using the internet'). As with any other source of information, that source has to be acknowledged through a citation in the text and a reference in the reference list. The citation in the text is, as for the other types of source, author(s) name(s) and date of publication. The basic format for the reference list is as follows:

Publication date

Article title (in italics)

Type of medium [in square brackets]

Names and initials of the author(s) or organization name.

Lunar Development Agency. 2018. *Mining the Lunar Surface* [online] Available from: http://www.lunardevelopment.co.uk/htm [Accessed 01 April 2018].

Date the site was accessed [in square brackets]

Web address

Digital object identifier (DOI)

The DOI is a means of identifying specific links to articles on the net. When an article is published and made available electronically, the publisher will give it a DOI number that can then be used to locate the article. When you draft your reference list it is good practice to include the DOI for print-version articles, as well as those that are only published on the web. An exemplar DOI is https://doi.org/10.1186/1471-2474-13-218, which will link through to a paper on peripheral muscle fatigue.

In this section, we have looked at the styles of citation and referencing for the most common sources of information that you might be using for your work. As stated before, there are different formats for presenting such information and, here, we have concentrated on the Harvard format, which is one of the standard forms used in the biosciences. You should, however, ensure that you check with your university or college to see if they have a preferred style of referencing.

 Chapter summary

This chapter has looked at the range of ways of working with sources to find information. This started with looking at the different types of sources available to you, ranging from textbooks through to research papers, and considering the type of information each source provides and the situations in which you might make use of them. We then reviewed the processes of searching for information, using a variety of online approaches, with a discussion of how to refine your searches and also how to assess their reliability. We considered the processes of making notes from these different sources and how to critically evaluate a research paper in order to be able to judge the quality and reliability of the content, and the way in which it was presented. In the final section of the chapter, we looked at the how to reference the material correctly in your written work.

Academic integrity and avoiding plagiarism

 Introduction

Academic integrity underpins all academic endeavours. This chapter explores what we mean by academic integrity and plagiarism, which is failure of academic integrity and is an increasing problem. We will begin by defining what plagiarism is and then look at some of the reasons students give for plagiarizing in their work. We will then outline a method for you to avoid plagiarism and highlight some of the many good reasons there are for making sure you do. It is important to note that, although most forms of plagiarism probably occur in the context of essays and practical reports, it is possible to plagiarize in any form of communication; so the study skills required to help you to avoid plagiarism in your writing are essential to your success in every area of your academic work.

5.1 What is academic integrity?

Academic integrity is a term that was developed by the American academic Don McCabe. In its broadest terms, it means being open and honest in all your academic work. In particular, this means giving due credit to those people whose ideas or work you use in your own work and being honest in presenting the outcomes of your research. This chapter focuses on the first aspect—giving due credit to the work of others—and its opposite, which is plagiarism.

Academic integrity means that whenever you make use of the ideas, writings, images, or other materials that have been created by other people, you should acknowledge that and not give the impression that these were your own. Not to give that acknowledgement is plagiarism, which is a significant problem in universities. This is partly due to the ease of access to information via the internet, and partly due to the copy-and-paste features of word-processing applications, making both finding and copying from one source to another a very easy process. In addition to this, the academic conventions surrounding plagiarism are not always widely understood; students often arrive at university with only a very limited understanding of the issue. Because of the rise in the incidences of plagiarism, universities are now much more aware of the matter and are, therefore, keen to address it. Many universities are now using plagiarism detection software to detect cases, which has made it much more difficult for students to get away with plagiarism. However, it is important not to get the issue out of perspective. As we shall see, plagiarism is often not deliberate and there are many positive reasons for learning how to avoid it, but first, we need to define what the term 'plagiarism' means.

5.1.1 Definition

The *Oxford English Dictionary* defines plagiarism as:

> The action or practice of taking someone else's work, idea, etc., and passing it off as one's own; literary theft.

So plagiarism is about using someone else's work, but giving the impression that it is your own. Let's think for a moment about the different elements of this definition:

- *taking*: could be from a book, journal, internet site, a lecture or handout, or even another student's work;
- *someone else's work*: could be someone else's words, but could also be their ideas, data, or images;
- *passing it off as one's own*: could be either deliberate or accidental, and could be in the context of an essay, a lab report, or a presentation (in fact, any assessed piece of work);
- *literary theft*: tells you, in unambiguous terms, how it is perceived by many, including your tutors.

So put simply, plagiarism is a form of cheating. You wouldn't cheat in exams and you shouldn't cheat in coursework either. Therefore, just as there are penalties for cheating in exams, there are penalties for cheating in coursework, too.

5.1.2 Good academic practice

Why are universities generally, and academics in particular, so concerned about plagiarism? Why is it considered to be such a big deal? Are they over-reacting or are there good reasons for their concern? Well, let's think for a minute about what underpins academic work. To be more specific, let's consider what is good academic practice.

It was Isaac Newton who said of himself (in a letter to Robert Hooke in February 1676):

> If I have seen further it is by standing on the shoulders of giants.

By 'have seen' he was referring to the tremendous discoveries he had made during his lifetime; by 'standing on the shoulders of giants' he was largely referring to the people who had gone before him whose ideas he developed and built upon. In making this statement Newton was highlighting a very important principle of academic study—that although you might research and then write up something individually, you are, in fact, contributing to a collective endeavour. Even the most novel or original of discoveries will be built upon the ideas of others, and it is important that those ideas are acknowledged and that it is only your own contribution that you claim for yourself.

So, plagiarism is an important issue because academic integrity underpins good academic practice—it is fundamental to all academic work. If you think about it, this also means that if you can learn the study skills that enable you not to plagiarize, you will be helping yourself not only to avoid being penalized, but more positively, to perform well in all areas of your academic work.

5.1.3 Deliberate plagiarism

When people hear the word 'plagiarism', they probably think of someone deliberately copying text without referencing it, but as we have seen from the definition, plagiarism is often much broader than simply copying text. We will deal first with examples of deliberate plagiarism.

Copying

Imagine that a student used the following text in an essay on 'Factors influencing the development of cancer'.

> There are many factors which influence the development of cancer. These include both endogenous factors, such as inherited predisposition, and exogenous factors, such as exposure to environmental carcinogens and infectious agents. Another factor which has a clear influence on the type of cancer which develops is age.

The first sentence is the student's own but the next two sentences are not written by the student; rather they are taken from the *Introduction to Cellular and Molecular Biology of Cancer* by M. A. Knowles and P. J. Selby (4th edn, 2005 OUP). The student has copied the text without any indication that some of the words have come from another source. Therefore, they have plagiarized by copying.

Alternatively, take an example of an image—imagine a student had used Figure 5.1 in an essay on 'Factors influencing the development of cancer'. If the student failed to acknowledge the source of the image, in this case Knowles and Selby (2005), Figure 1.3, p. 8, they would also be plagiarizing by copying. It is important to remember that plagiarizing by copying images is equivalent to, and so just as serious as, plagiarizing by copying of text.

Figure 5.1 Sample of a plagiarized image

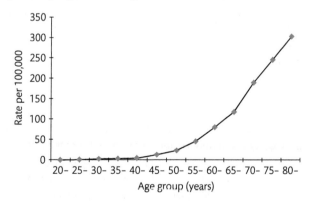

Taken from Knowles and Selby (2005), Figure 1.3, p. 8 .

Collaboration and collusion

The boundary between collaboration and collusion is one that students often find confusing. At one end of the spectrum, you may be required to collaborate in group work exercises, in which a single piece of work is produced, representing the combined efforts of the group. Often such pieces of work may be awarded a single mark that is given to all the members of the group. At the other end of the spectrum, two students may work together on a piece of work that is supposed to be an individual exercise. The latter is an example of collusion since the assignment is supposed to reflect the work of the individual students working independently. Under those circumstances, the students would normally be subject to the same penalties as for plagiarism. Likewise, if you knowingly allow another student to copy your work, you will also be in breach of your university's regulations. In this case, remember that this also applies if the student is in a different year, for example, the year below you, or is studying at another university. Because plagiarism checking systems such as Turnitin™ keep a database of all submitted work, there is automatic cross-checking between year groups and institutions.

The boundary becomes more blurred where part of the activity requires collaboration, but the final product has to be an individual effort. The typical example of this would be working with a partner in a practical class. It is important to remember that while you will often need to work with a fellow student to conduct a practical experiment or undertake fieldwork, this doesn't usually mean that you are being asked to collaborate on the writing-up aspect of the report as well. Obviously, there will be similarities between two reports written by two students who had worked together to conduct an experiment, but these should largely be confined to the methods and results sections. However, in the analysis of the results and in the discussion section, the students will be expected to analyse the data and draw conclusions from their findings *independently* of each other. For example, compare the two discussion sections in Figure 5.2 from students A and B, who were lab partners for a practical class on electrical activity in skeletal muscles.

You can see that there are definite similarities between the two discussions; the marker would probably conclude that the students had collaborated on the write-up of the discussion section and had, therefore, plagiarized.

Now look at the examples of students A and C in Figure 5.3.

Although we would still expect to see similarities in the methods and results sections, we can see that the students have thought independently about their findings and have come to their own conclusions. This doesn't mean (in this case) that they have come to *different* conclusions, but the wording and sentence structure suggests that they have thought the ideas through for themselves.

Self-plagiarism

It is also possible to self-plagiarize. This is when students submit a piece of their own work, which they have previously submitted for another part of their course, but fail to declare that it had been used for such a purpose. This is probably the type of deliberate plagiarism that students find the most frustrating; 'I'm the author of the work' you might reason, 'so surely I can submit it as my own work for whatever aspect of my course I choose.' It can be difficult not to have some sympathy with this view, but nevertheless self-plagiarism is considered

Figure 5.2 Collaboration in lab write-ups

Student A	Student B
Discussion	**Discussion**
In these experiments, observations were made of the changing patterns of electrical activity in skeletal muscles during voluntary contractions. During contractions of progressively increasing force, there was an increase in the electrical activity recorded from the muscle. This increase was related to the two different ways in which the nervous system controls muscle force: an increase in the action potential frequency of individual motor units and the recruitment of additional motor units (Berne and Levy, 2000).	Recordings have been made of the patterns of neural activity in skeletal muscles during the course of voluntary contractions. When the force was progressively increased, there was an increase in the electrical activity recorded from the muscle. This increase was attributed to the two ways the nervous system regulates muscle force, namely by increasing the action potential frequency of individual motor units and by recruiting additional motor units (Berne and Levy, 2000).
Observations of single unit activity showed that the firing began at about 8 impulses/s, increasing to 40 impulses/s at maximum force . . .	Recordings of the activity in single motor units showed that the discharge began at about 8 impulses/s and increased to 40 impulses/s at the maximum force . . .

Figure 5.3 Independent lab write-ups

Student A	Student C
Discussion	**Discussion**
In these experiments, observations were made of the changing patterns of electrical activity in skeletal muscles during voluntary contractions. During contractions of progressively increasing force, there was an increase in the electrical activity recorded from the muscle. This increase was related to the two different ways in which the nervous system controls muscle force: an increase in the action potential frequency of individual motor units and the recruitment of additional motor units (Berne and Levy, 2000).	The aim of the experiments was to use electromyographic (EMG) recordings from the whole hypothenar and from single motor units to investigate the neural control of muscle force during voluntary contractions. The initial test contraction involved a controlled increase in force and the recordings showed a progressive increase in the level of EMG activity paralleling that of the force production. It was predicted that this increase in EMG activity was made up of two different elements of neural control: variation in the rate of action potential discharge in the individual motor units and recruitment of additional motor units (Berne and Levy, 2000), these two processes occurring in parallel.
Observations of single unit activity showed that the firing began at about 8 impulses/s, increasing to 40 impulses/s at maximum force . . .	To test this prediction, recordings were made of the activity of single motor units . . .

unethical because of the deception involved in submitting the same material for credit in different modules or courses. On a more positive note, writing can always be improved upon, so it is in your own interest to do more than just recycle previously submitted work; it also helps the learning process. This is in addition to the more obvious point that a good answer written for one essay question, for example, is probably not going to be specific enough to be considered a good answer to a different, even if related, essay question. In short, although disallowing self-plagiarism may seem a little harsh, it is to your advantage to tackle each piece of assessed work as a separate exercise; allowing you to tailor your answer to the specific task and broaden your knowledge of the subject.

Buying

There are increasing numbers of websites, often referred to as 'essay mills', through which it is possible to buy essays, even complete dissertations, for submission as assessed work. This is the most blatant and outrageous form of plagiarism there is. Just in case you are ever tempted to do this, here are a number of good reasons why you shouldn't:

- it's expensive;
- it's fraudulent;
- you learn nothing about the subject;
- it will be easy for your tutors to spot if there is a significant disparity between a bought essay and your usual standard of work;
- the penalties, if you are found out are likely to be very severe—often expulsion from the university.

Don't do it!

5.1.4 Accidental plagiarism

Some plagiarism is deliberate, but more commonly it is accidental, due to poor academic practice. There are lots of reasons for accidental plagiarism; we deal with some of them in the following paragraphs, and use them to form the basis for Section 5.2, 'How to avoid plagiarism'.

Ignorance

Not that you will have this excuse after reading this chapter, but ignorance is the most basic excuse. 'I didn't know' is a common response to accusations of plagiarism, particularly from undergraduates in the early stages of their degree course.

Poor planning

Poor planning or failing to manage your time well may seem unrelated to plagiarism, but they are, in fact, frequent causes of it. Most students would be shocked at the idea of deliberate plagiarism, especially in a pre-planned manner, but when you are running out of time to complete an essay the temptation to take short-cuts becomes stronger. Read the case study that follows and see if you can appreciate what we mean.

Case study: Steve

> Steve was busy with another assignment and then went out to celebrate when it was finished. Waking up in the morning he remembered that he had a second assignment to hand in and so quickly used Google to search for some information on the web. Using a mixture of copying, pasting, and rephrasing he put together his assignment in a couple of hours. However, his tutors spotted what he had done and he was given a mark of zero.

So plagiarism as a result of poor planning is more plausible than it may at first seem.

Failure to record source details

Failure to record source details is probably the simplest form of plagiarism. It starts at the note-making stage (as we referred to in Section 4.3, 'Note-making strategies'); you find an appropriate piece of text, image, or data that you would like to reproduce or adapt in some way for your essay or report, but you don't write down where the material came from. At the assignment-drafting stage you either forget you got it from somewhere else and believe it to be your own original work, or you realize it came from elsewhere, but either can't be bothered or don't have the time to retrace the source, so you use it anyway without referencing and hope that no one will notice. Does this seem implausible? Read the following case study and see if you understand what we mean.

Case study: Andria

> Andria took a lot of notes while she was researching her assignment, but didn't pay much attention to noting down the referencing details of her sources. When she wrote her assignment she ended up very unsure about what were her own ideas and what she had taken from books. Her lecturer is now talking about plagiarism, but she feels that she was just a bit disorganized and is horrified that she is being accused of dishonesty.

Plagiarism due to failure to record source details is, therefore, a very real problem.

Inappropriate notes

Inappropriate notes are also a common cause of plagiarism. It is reasonably straightforward to record source details (in theory at least), but how do you know what sort of notes to make on the source? See if the following case study helps you understand what we mean.

Case study: Abdul

> Abdul has a good understanding of the course, but he finds it difficult to put things into his own words. Academic textbooks always seem to put things so well that he hasn't got much to add. Sometimes it just seems easier to copy out chunks of different books, paraphrase them a bit, and organize them into an answer. He doesn't feel that this should be considered plagiarism as he has researched the material and put the assignment together from a range of different sources. Unfortunately his tutors don't agree.

Academic textbooks or journals do, indeed, 'put things so well' that it's difficult for a student to know what to add. This is why it is so important to work at understanding the concept that you are dealing with, and it takes a certain amount of confidence to do this. We will deal with this in Section 5.2, 'How to avoid plagiarism'.

Incomplete citing and/or referencing

Sometimes students inadvertently plagiarize by simply not recording their citations and references fully. This could be due to a lack of understanding of what is required, or a lack of willingness to add what are sometimes just considered to be 'finishing touches', but are, in fact, crucial to the integrity of their work. Again, see if the following case study helps to illustrate this.

Case study: Jasmine

> Jasmine worked hard on her essay and was pleased with what she produced. However, she has never understood the point of the referencing system. She included all of the books that she used in a reference list at the end of the essay, but didn't bother citing her sources in the body of the essay. She thought the reference list at the end would be enough, but unfortunately, her lecturers viewed this lack of citing as plagiarism. They covered the assignment in comments and gave her a very low mark.

So, although sorting out the citations and references might just seem like an optional extra, it is vital to get it right. Referencing correctly is dealt with in Section 5.2.5, 'Reference correctly', and more detailed guidance is available in Section 4.5, 'Citations and references'.

Lack of engagement with sources and failure to express own ideas

This may possibly be the reason that requires the most thinking to address. It is where a student may well have planned ahead, recorded source details, made appropriate notes, and cited and referenced accurately, but still doesn't demonstrate that they have thought a great deal about what they have written. It is perfectly possible, in an assessed piece of work, simply to string together lots of other authors' ideas (either quoted or paraphrased), but if you fail to include your own ideas, in the form of comments or analysis of the material, you still might be demonstrating poor academic practice, even if the source material is cited and referenced correctly.

5.2 How to avoid plagiarism

We have dealt with what plagiarism is, having thought about its definition and how academic integrity (or not plagiarizing) underpins good academic practice. We have also considered the main reasons students plagiarize, identifying both deliberate and accidental forms of plagiarism. We are now going to consider how to avoid it.

5.2.1 **Know what plagiarism is**

The first step to avoiding plagiarism is to be able to recognize it when you see it. This is different from simply defining it; can you tell in practice when something is or is not plagiarized?

Try this—Essay extract exercise[*]

Original text

> During the last 60 years the development of effective and safe drugs to deal with bacterial infections has revolutionized medical treatment, and the morbidity and mortality from microbial disease have been dramatically reduced (Rang *et al.*, 1999).

Essay extract 1

> During the last 60 years the development of effective and safe drugs to deal with bacterial infections has revolutionized medical treatment, and the morbidity and mortality from microbial disease have been dramatically reduced.

Is this extract plagiarized, yes or no?

Essay extract 2

> During the last 60 years the development of effective and safe drugs to deal with bacterial infections has revolutionized medical treatment, and the morbidity and mortality from microbial disease have been dramatically reduced (Rang *et al.*, 1999).

Is this extract plagiarized, yes or no?

Essay extract 3

> 'During the last 60 years the development of effective and safe drugs to deal with bacterial infections has revolutionized medical treatment, and the morbidity and mortality from microbial disease have been dramatically reduced' (Rang *et al.*, 1999).

Is this extract plagiarized, yes or no?

Essay extract 4

> In the 4th edition of their textbook *Pharmacology* (1999), Rang, Dale and Ritter state that: 'During the last 60 years the development of effective and safe drugs to deal

---

[*] Exercise adapted from Willmott, C.J.R. and Harrison, T.M. (2003). An exercise to teach bioscience students about plagiarism. *Journal of Biological Education* **37**, 139–140. (Thanks to the *Journal of Biological Education* for permission to adapt this article.)

(Continued)

with bacterial infections has revolutionized medical treatment, and the morbidity and mortality from microbial disease have been dramatically reduced.' Such a bold assertion understates the ongoing threat posed by microbial infection. It is estimated, for example, that worldwide there were over 8 million cases of tuberculosis in 1998 (WHO, 2000).

Is this extract plagiarized, yes or no?

Essay extract 5

The development of safe and effective drugs to deal with bacterial infections has dramatically reduced the death rate arising from microbial diseases (Rang *et al.*, 1999).

Is this extract plagiarized, yes or no?

Essay extract 6

During the post-war years, the development of effective and safe drugs to deal with bacterial infections has transformed medical treatment, and death and illness resulting from microbial disease has been dramatically reduced (Rang *et al.*, 1999).

Is this extract plagiarized, yes or no?

Essay extract 7

The availability of antimicrobial compounds has transformed healthcare in the period since the Second World War. People are far less likely to die or even be seriously ill than they had been prior to the introduction of these drugs (Rang *et al.*, 1999).

Is this extract plagiarized, yes or no?

Essay extract feedback

Essay extract 1

This is copied word-for-word, but there is no reference. It is a definite case of plagiarism.

Essay extract 2

This is marginally better than extract 1 because the source has been acknowledged by the inclusion of the reference. However, it is still plagiarism: the text has been copied word-for-word and should be in quotation marks to indicate that it is the exact wording of the source.

Essay extract 3

Quotation marks are used to acknowledge that words have come from a different source. The quotation marks here make it clear that the student is acknowledging that both the ideas and the words have come from the textbook, so it is not plagiarized. However, stringing together a series of quotations does not demonstrate your understanding of the subject and so is likely to score low marks (this will be dealt with in Section 5.2.6, 'Use sources appropriately and express own views').

Essay extract 4

The quotation is indicated and used in an appropriate way; it has been commented on by the author and supplemented with further information, and so is not plagiarized.

Essay extract 5

This is only a cosmetic alteration. The wording and sentence construction very closely resembles the source and so is plagiarized.

Essay extract 6

This student has just swapped a few words, perhaps using a thesaurus to find replacements, but this is not sufficient to be considered new work. The thinking has gone into finding alternative words to avoid directly quoting, rather than into understanding the statement, and so is plagiarized.

Essay extract 7

The student has understood the source and put the relevant information into his or her own words. This demonstrates the student's engagement with the text and ability to explain the information relevant to the essay, and so is not plagiarized.

5.2.2 Plan ahead

We have already noted that bad planning, and the pressures it creates, increases the temptation to plagiarize. So while bad planning is not a direct cause of plagiarism, it is still a significant factor, and so needs addressing.

The solution to bad planning is, of course, good planning. This strikingly obvious statement is worth mentioning, as it highlights the fact that the skills required to address this issue are simply common sense. However, there can be a world of difference between knowing what you should do and actually doing it. In Chapter 7, 'Writing essays', we deal with the different stages involved in writing an essay. It is helpful to think of these stages not just in terms of separate tasks, but also as distinct periods of time that need planning. Chapter 7 recommends that you think about the following stages when writing an essay:

- analysing the question and brainstorm ideas;
- researching the topic;
- writeing a plan;
- writeing a first draft;
- reviewing and redrafting;
- proofreading (including checking of citations and references).

A good way to plan ahead is to order these tasks into some sort of timeline, calendar application, or diary. Figure 5.4 shows an example timeline representing a period of three weeks

Figure 5.4 Essay tasks timeline

that you might have available to write an essay. It is worth spending a short amount of time allocating your tasks in this way to provide some structure and discipline to your time, and so hopefully reduce the chances of you leaving things to the last minute, which would, as we have seen, increase the chances of you plagiarizing.

This method also has the advantage of making the task less daunting. If you had on your 'to do' list 'write ion channel essay', it would probably be a long time before you were able to cross it off, which can be quite demotivating. However, if your 'to do' list said things like 'gather journal information for essay' and 'write essay plan', these are things that would take hours, rather than weeks, so you would be able to cross things off much sooner. This makes it easier to discern progress and so is motivating, rather than demotivating; you feel like you are getting somewhere.

Note also in Figure 5.4 that there is a significant gap between 'Write first draft' and 'Review and redraft'. This is important for a couple of reasons: first, you could reward yourself in some way for getting this far in the process, again a motivating factor; and secondly, coming back to an essay after a bit of a break enables you to be more objective about what you have written. This objectivity increases the likelihood that you will be able to make significant improvements.

5.2.3 Record source details

As we have seen, it is important to record source details if you don't want to plagiarize. It sounds simple enough, but as with planning ahead, when the pressure is on, it is both easier and more tempting to make mistakes.

Imagine you had made this note during a study session in the library last week:

> Ion channels are not simply tubes that span membranes through which ions can rapidly flow. Instead, they are highly sophisticated molecular machines that respond to chemical and physical changes in their environments and undergo precisely timed conformational changes.

Table 5.1 Source details to record for different information types

Source type	Source details required
Books	Name(s) of author(s)*
	Year of publication
	Title of book
	Name of publisher
	Place of publication
	Page reference
Journals	Name(s) of author(s)
	Year of publication
	Title of paper
	Title of journal
	Volume number
	First and last page number
Websites	Name(s) of author(s)
	http://address
	Title
	Date page accessed†
Lectures	Name of lecturer
	Name of lecture/lecture series
	Date of lecture
	Location (e.g. University of Leicester)

*If it's a chapter of an edited book you also need to include the 'Editor(s) name(s)', 'Title of chapter' and 'First and last page numbers of chapter'.

†Often internet pages indicate when they were last updated—this is also important information to record if available.

It is a good piece of writing and perfect for your essay on ion channel function. The only trouble is you can't quite remember if these are your words or someone else's because you didn't write down any more information; an easy mistake to make.

The words are actually taken from *Biochemistry* by Berg *et al.* (5th edn, 2002, WH Freeman & Co. Ltd). So if you were to use them in your essay without putting them in quotation marks ('...'), citing, and referencing them, you would be plagiarizing. The good news is that, because failure to record source details is a simple mistake to make, there is also a simple solution. It just takes a bit of discipline.

There are four main sources that you will use in your studies: books, journals, websites, and lectures. Table 5.1 summarizes the information you need to record for each one in order to reference fully.

All you need to do is make sure you record the source details relevant to the type of source before you begin to make notes. It's as simple as that. See Table 5.2 for examples of citations and references from the source details.

Table 5.2 Examples of source details recorded

Source type	Source details required	Example source details	Example citation	Example reference
Books	Name(s) of author(s): Year of publication: Title of book: Name of publisher: Place of publication: Page reference:	Dawn Hawkins 2005 Biomeasurement OUP New York pp. 95–96	Hawkins (2005)	Hawkins, D. 2005. *Biomeasurement*, OUP, New York, pp. 95–96.
Journals	Name(s) of author(s): Year of publication: Title of paper: Title of journal: Volume number: First and last page number:	Watson, J.D. and Crick, F.H.C. 1953 Molecular structure of nucleic acids; a structure for deoxyribose nucleic acid Nature 171 737–738	Watson and Crick (1953)	Watson, J.D. and Crick, F.H.C. 1953. Molecular structure of nucleic acids; a structure for deoxyribose nucleic acid. Nature 171, 737–738.
Websites	Name(s) of author(s):	Heitler, W.J.	Heitler (2008)	Heitler, W.J. 2008. How grasshoppers jump. [online] available from: http://www.standrews.ac.uk/~wjh/jumping[accessed 27 June 2008].
Lectures	Name of lecturer: Name of lecture/series: Date of lecture: Location:	Jon Scott Physiology of locust hearing 23/11/17 University of Leicester	Scott (2006)	Scott, J. 2017. Physiology of locust hearing [lecture 23 November 2017].

5.2.4 **Make appropriate notes**

Recording source details, then, is quite straightforward, as long as you know what details you need to record for which source. However, what about when you actually need to make notes on the source material? How do you do it?

When making notes it is important to ask yourself a simple question: what am I making notes for? There are several possible reasons you might have. You might be making notes to:

- select information;
- understand information;
- remember information.

Clearly these three areas overlap considerably, but in order to keep things simple we are going to deal with them separately. Chapter 4, 'Working with different information sources', deals with making notes to understand information and the reading techniques necessary to achieve this efficiently; Chapter 13, 'Getting the most out of revision', deals with taking notes to remember information. Here, we will just focus on taking notes to *select* information.

At the beginning of this chapter, when we were thinking about the different elements involved in the definition of plagiarism, we identified that 'someone else's work' could be their words, but also their ideas, data, or images. When making notes, therefore, we are not merely thinking about text; we also need to remember other forms of information, too, including ideas, data, and images. In simple terms, you have two options when considering how to include another person's work in your own: you could reproduce it exactly (so in the case of text this would mean quoting it word-for-word), or you could adapt it in some way (again, in the case of text this would mean putting it into your own words, i.e. paraphrasing it).

For example, if you were writing an essay on 'Factors influencing the development of cancer' and found some useful information in *Introduction to the Cellular and Molecular Biology of Cancer* by Knowles and Selby (4th edn, 2005, OUP), you may decide that the exact wording and/or imagery is important to you and so want to reproduce it exactly, as shown in the example which follows and in Figure 5.5.

Text

> 'There are many factors which influence the development of cancer. These include both endogenous factors such as inherited predisposition and exogenous factors such as exposure to environmental carcinogens and infectious agents. Another factor which has a clear influence on the type of cancer which develops is age' (Knowles and Selby, 2005).

Image

However, if it was the *idea*, rather than the exact wording or imagery that was important, you would need to adapt it in some way (which may mean simplifying or expanding on it), as as shown in the example which follows and Figure 5.6:

Figure 5.5 Colon cancer incidence in men in England in 1999

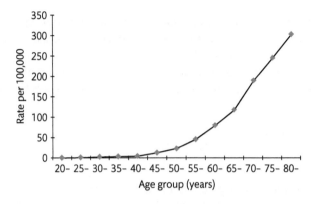

Reproduced from Knowles, M. & Selby, P. (2005). *Introduction to the Cellular and Molecular Biology of Cancer*, with permission from Oxford: Oxford University Press.

Figure 5.6 Colon cancer incidence in men in England in 1999

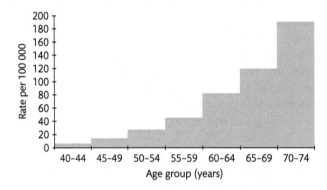

Adapted from Knowles, M. & Selby, P. (2005). *Introduction to the Cellular and Molecular Biology of Cancer*, with permission from Oxford University Press

Text

> There are many reasons for the development of cancer, but most can be grouped into two broad categories; those originating from within the organism (e.g. genetic predisposition) and those having an external origin (e.g. carcinogens). A third, but no less important, category is age (Knowles and Selby, 2005).

Image

The important thing to remember is that whether you are reproducing another's work (e.g. quoting text) or adapting it (e.g. paraphrasing text), both need referencing. In the case of quoting, the quoted text is placed in quotation marks ('…'; to make clear that you are using someone else's words verbatim) followed by the citation. In the case of paraphrasing the paraphrased text is followed by the citation.

5.2.5 **Reference correctly**

As with planning ahead and recording source details, referencing correctly is simple enough in theory, you just need to pay attention to the details. Here are some simple steps that should help (more detailed guidance is available in Section 4.5, 'Citations and references').

Check what style of referencing is expected from you

Different institutions have different styles of referencing; the most common system in the biosciences is an author–date system (see Section 4.5), but some will use footnotes, some will use endnotes, some will use variants of these systems, and there are many others besides! Even within an individual institution it is unlikely that all departments will use the same type of referencing style. In fact, there are likely to be minor variations even within departments. This can create a lot of confusion and it can be very tempting to decide not to bother with the details. However, when it comes to referencing, the details are important.

Know the difference between citing and referencing

An important distinction to make is the difference between citing and referencing. In general terms *referencing* refers to the practice of acknowledging your sources. More precisely, however, referencing in an academic context has two elements:

- the citation in the text;
- the full reference at the end of the piece of work (e.g. your essay or practical report).

The *citation* is simply the way you refer to another's work in your own work; in the case of the author–date system this means the text, idea, image, or data followed by the surname of the author (or authors) and the date of the publication. For example, 'The novel feature of the structure is the manner in which the two chains are held together' (Watson and Crick, 1953). The *reference* is the full details of the authors and publication at the end, for example Watson, J.D. and Crick, F.H.C. 1953. Molecular structure of nucleic acids; a structure for deoxyribose nucleic acid. *Nature*, **171**, 737–738 (see Figure 5.7).

Figure 5.7 The relationship between citations and references

'The novel feature of the structure is the manner in which the two chains are held together'
(Watson and Crick, 1953).

References

Watson, J.D. and Crick F.H.C. 1953. Molecular structure of nucleic acids; a
   structure for deoxyribose nucleic acid. *Nature* **171**, 737–738.

Use *et al.* correctly

Et alii is a Latin term meaning 'and others'. It should, therefore, be written in italics (because it's Latin) and be followed by a full stop (because it's an abbreviation). It is used when you are citing a source that has three or more authors, for example, when referring to the book *Molecular Biology of the Cell* published by Garland Science, instead of having to write Alberts, B., Johnson, A., Lewis, J., Raff, M., Roberts, K., Walter, P. (2002), you would simply write, Alberts *et al.* (2002), which, as you can see, is very handy.

List the references in alphabetical order at the end

Your reference list at the end of your document should be in alphabetical order by surname of the author, for example:

> Dalgleish, G. 2012 . . .
>
> Harrison, W. 2015 . . .
>
> Wells, B. 2014 . . .

If there is more than one author for a publication then it is the surname of the first author that counts, for example:

> Cann, T. 2015 . . .
>
> Jones, M. and Adams, C. 2013 . . .

If there are two authors with the same name or one author with two publications, then you need also to order the list by date, for example:

> Dalgleish, G. 2012 . . .
>
> Dalgleish, G. 2014 . . .

If an author has more than one publication in a given year, then the list needs to distinguish between these, usually by including letters in the dates, for example:

> Robbins, H. 2014a . . .
>
> Robbins, H. 2014b . . .

This distinction would also need to be reflected in the citation, so (Robbins, 2014a), and so on.

Consider using bibliographic software, if available

If you are going to be using a large number of references, for instance in a final year dissertation, you may wish to use bibliographic software. Bibliographic software is designed to allow you to input, organize, manage, retrieve, and format lists of references more easily than manual input. Two examples are EndNote and RefWorks, but it will depend on what software your university subscribes to. Whatever system your university subscribes to, it is worth finding out more about it as it may save you a lot of time. However, even if you use such software, you still need to understand the structure of references if you are going to use it appropriately.

5.2.6 Use sources appropriately and express your own views

We noted in Section 5.2.4, 'Make appropriate notes' that, 'In simple terms you have two options when considering how to include another person's work in your own'—reproducing exactly or adapting for your own purposes. These are only simple terms because merely using another person's work and referencing it correctly is only a starting point. As essay extract 3 (Section 5.2.1, 'Know what plagiarism is') illustrated, if an essay or report were just a collection of other people's ideas it wouldn't score very good marks; in order to do well you need to engage actively with source material and express your own views by providing comment or analysis on your sources. This is what this final section in 'How to avoid plagiarism' is about.

There are many ways in which you can engage with source material. We are going to look at three examples. Engaging with source material to:

- illustrate a point;
- provide evidence;
- contrast arguments.

Precisely which method of engagement you choose is somewhat secondary; the important point is that you must engage with the sources you use in some way, rather than using material without comment or analysis. (Thanks to Dr Raymond Dalgleish from the Department of Genetics, University of Leicester, for these examples.)

Illustrating a point

One way to use source material is to use it to illustrate a point. See Figure 5.8 for an example from the field of genetics. In this example, the idea is introduced, the evidence is summarized, and then an explanation is given of what this evidence illustrates.

Figure 5.8 Using sources to illustrate a point

More than four decades ago Marshall Urist began a series of experiments designed to characterize the factors necessary to induce bone formation (Urist, 1965). In 1975 he demonstrated that the 'bone morphogenetic factor' was part of the protein component of bone, rather than being associated with the hard mineralized fraction (Urist *et al.*, 1979). However, he was initially unable to isolate the factor as a single protein. This illustrated that the induction of bone growth might be controlled by several co-acting protein factors.

Introduction

Summary of evidence

Explanation of what this illustrates

Figure 5.9 Using sources to provide evidence

> Over several years, Urist continued to characterize his bone morphogenetic factor, making improvements to the purification procedures (Urist *et al.*, 1984). He showed that demineralized bone could be fractionated to yield low molecular weight protein morphogenetic components, but none of these was capable alone of inducing bone formation (Urist *et al.*, 1987). These studies suggest that bone morphogenesis is a complex process requiring the interaction of several factors.

Introduction

Summary of evidence

Explanation of evidence

Providing evidence

Another way to use source material is to use it to provide evidence. See Figure 5.9 for an example, again from the field of genetics. In this example, the idea is introduced, the evidence is summarized, and then the evidence is explained.

Contrasting arguments

A third way to use source material is to use it to contrast arguments. See Figure 5.10 for an example, again from the field of genetics. In this example, the idea is introduced, some evidence is paraphrased, this evidence is criticized, contrasting evidence is then introduced, and finally the writer's opinion on the issue is offered.

Figure 5.10 Using sources to contrast arguments

> For many years, the precise identity of Urist's bone morphogenetic protein (BMP) remained unresolved except for the fact that it had a major component with a molecular weight of 18 kDa. In 2005 Behnam *et al.* demonstrated that the amino acid sequence of peptides derived from highly purified BMP show identity to regions of a previously characterized protein known as spp24. However, spp24 has a molecular weight of 24 kDa. It could be argued that Behnam *et al.* reached a mistaken conclusion because of failures in the strategy used to characterize BMP or that they simply failed to fully explain their findings in the published account of their work. Subsequent experiments, involving recombinant protein expression, have shown that the spp24 protein undergoes post-translational cleavage, yielding a product consistent with Urist's original 18 kDa BMP.

Introduction

Paraphrasing

Criticism

Speculation about the discrepancy

Resolution of the discrepancy

These three ways of engaging with source material are not exhaustive; there are many others including agreeing, disagreeing, synthesizing, reconciling, and developing. The important thing is that you provide some comment or analysis on the sources you use, rather than simply recording them.

5.3 Six good reasons not to plagiarize

As we have seen, avoiding plagiarism is not just about correct references, it is fundamental to good academic practice. In case you need any more convincing, here are six good reasons for you not to plagiarize.

1. *Plagiarism short-circuits learning:* When you are set an assessed piece of work it is easy to focus on the tangible endpoint, that is, the mark you are awarded. However, your tutors set you work, not merely as a means of allocating marks, but also as a means of helping you to learn. The research you conduct, the selecting of information, and the process of writing all aid the learning process. The more you plagiarize, the less you learn.

2. *Plagiarism destroys independent thinking:* Your tutors are interested in what you think about a subject. Obviously, your point of view needs to be supported by appropriate evidence, but, as we have seen, that evidence needs to be accompanied by your own comment or analysis; your own ideas and conclusions. If you plagiarize by not commenting or analysing evidence, you are not thinking for yourself; you are merely communicating the ideas of others and have failed to think independently.

3. *Plagiarism is unfair to your peers:* How would you feel if a student on your course scored a higher mark than you on an essay, but you knew that they had cheated? Make sure you don't make anyone else feel the same way about you.

4. *Plagiarism is a form of theft:* Words, ideas, data, and images that have been written, thought of, discovered, or created by someone else belong to that person. Using them without acknowledging the source is stealing.

5. *Plagiarism can result in severe penalties:* Different universities have different penalties for students who have plagiarized. The type of penalty depends on the precise nature of the plagiarism, the extent to which the work contributes to final marks, and whether or not it is a first offence. Common penalties include:

   - an awarding of zero for the piece of work, the requirement to repeat the work, plus an official warning;
   - an awarding of zero for the module and withdrawal of the right to resit;
   - the downgrading of degree class by one division;
   - expulsion from the course.

6. *Plagiarism is now more difficult to get away with:* Many universities are now using plagiarism-detection software. Students are required to submit their work electronically, and the software then compares the work with a vast array of other sources, including core texts, journal articles, internet sites, and previously submitted student work. A report is produced highlighting the areas of similarity with the compared sources (Figure 5.11).

Figure 5.11 Example of a plagiarism software originality report generated by Turnitin[TM]

Reproduced with permission from Turnitin International.

The tutor would then interpret the report to decide whether or not it is plagiarized. The software is becoming increasingly sophisticated and makes getting away with plagiarism much more difficult. However, even if your university doesn't currently operate an electronic detection system, it is worth noting that academics are also very capable of quickly spotting plagiarism using more old-fashioned methods.

Finally, remember that avoiding plagiarizing is not simply about evading punishment. The skills that help you not to plagiarize are the skills that enable you to be an effective learner. Learning these skills will help you in all areas of your academic work, helping you to understand your subject better and so helping to improve your grades.

Chapter summary

As we have seen, academic integrity and good academic practice are really important. Failure to adopt these approaches can lead to plagiarism so you need to know both what plagiarism is and how to avoid it if you are going to do well in your studies. Planning, recording source details, making appropriate notes, referencing correctly, using sources appropriately, and expressing your own views are all important elements. If you can address these issues you won't simply be avoiding plagiarism, you will become a better scholar.

6 Choosing the right writing style

 Introduction

Writing for scientific purposes is often very different from the styles of writing we may use for other forms of communication, as it normally requires the use of a formal, impersonal style. It is important to use words accurately and to avoid slang, informal phrases, and generalities. When we are talking with each other, we can use gestures to add to the meaning of what we say, and we can repeat things or state them differently if we realize the listener has not understood us. It is also very easy to be rather lazy in our expression because we can see whether our listener has understood, '. . . know what I mean?' When you are writing, you have only the written words to convey your meaning and you cannot ask the reader whether they have understood or not, so precision is very important.

6.1 Voice

The style for scientific writing is normally impersonal and so avoids the use of personal pronouns such as 'I' and 'we' and the expression of subjective viewpoints. It also commonly involves using the *passive* voice rather than the *active* voice. For example, an introduction to your essay could be written as:

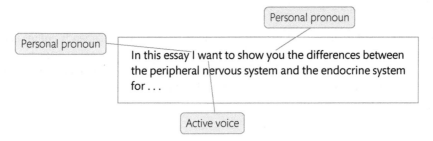

For a formal, scientific style it would be better to write:

Passive voice

In this essay, comparisons will be drawn between the peripheral nervous system and the endocrine system . . .

You should also avoid using contractions of words and expressing a subjective viewpoint. For example, the conclusion to an essay might be written:

> The evidence I've given supports my feeling that . . .

This phrase contains the personal pronoun 'I' in the contraction 'I've'. It may be that, for particular reasons of emphasis, you wish to keep the word 'I' in this phrase, but you should avoid using any of the many contractions that appear in conversation such as 'I've', 'don't', 'couldn't', etc. Instead you should write them out in full: 'I have', 'do not', 'could not'. [In this book we have chosen to use contractions because we feel that a more informal style is appropriate here, but it's (there we go again!) not appropriate for an academic essay.]

The phrase also expresses a very personal, subjective viewpoint: 'my feeling'. In the formal voice, this could be written as:

> The evidence presented supports the argument that . . .

If it is important to express a personal summing up of the arguments presented, then you could write:

> On the basis of the evidence presented, I would conclude that . . .

The use of generalizations often clouds the meaning of what is written or can indicate a lack of confidence on the part of the writer. For example, what does the following phrase actually mean?

> The peripheral nervous system quite often responds to stimuli . . .

The phrase 'quite often' has no real meaning as its interpretation depends entirely on the reader. Relative terms such as 'quite' or 'fairly', or phrases such as 'In general terms . . .', lack precision and should, wherever possible, be replaced with more precise phrases. Indeed, if you can, you should provide quantification of the statement, so rather than stating 'quite often' you can write:

> In 75% of cases the response was . . .

As with the choice of words, precision in grammar is very important to ensure that our writing is clear and unambiguous. For example, the incorrect position of a comma in a sentence can change the whole meaning of that sentence as in the old joke about the panda that walks into a bar, eats something, pulls out a gun and shoots the bar tender before leaving the bar. When questioned, the panda pulls out a book and points to the description of the panda which states it '. . . eats, shoots and leaves'. Had the description been phrased correctly as '. . .

eats shoots and leaves', the bar tender would not have been shot! There are numerous, entire books covering the subject of grammar (e.g. *Eats, Shoots and Leaves*, by Lynne Truss; Harper Collins, 2009) so in this brief section we will look at just a few of the key rules.

Having read this section you should now be well aware that we have not written this book in a scientific style!

6.2 Sentences and phrases

The fundamental unit of writing is the sentence. Sentences may be short. If too many short sentences follow on from each other, however, the style of writing can seem very abrupt and disjointed, but overly long sentences, particularly if they contain several, interlinked phrases, that might be associated with different points, can also be problematic because they can cause the reader to lose the thread of the argument being presented and give rise to the impression that the writer does not have a clear perspective on what he or she is trying to say.

The section you have just waded through is clearly badly written, and you may well not be sure what we were trying to say (as a short exercise, you could try rephrasing the paragraph so that the meaning is clear). So what is a sentence? At its most fundamental, a sentence comprises the following features:

For example:

In fact, this sentence contains two nouns: 'Neurons' and 'cells'. The noun 'Neurons' is the *subject* of the sentence, and the noun 'cells' is used to give the reader more information about what neurons are. Although it is very brief, it is a complete sentence, whereas:

Neurons are cells that are located

is a phrase, not a complete sentence. Although it contains a noun and a verb, the sentence is incomplete because the verb 'located' needs to be qualified by a description of where that location is:

> Neurons are cells that are located in the nervous system.

is now a complete sentence. We may also include *adjectives* and *adverbs* in order to give greater detail to our descriptions to the reader (see Box 6.1). Adjectives are words used to describe a noun and adverbs are used to describe a verb. For example, the following sentence:

> The hormone, testosterone, is active in the body.

is a complete sentence, but it does not give the reader very much information. By using adjectives and adverbs we can amplify the statement as follows:

> The steroid hormone, testosterone, is active widely in the body.
>
> Adjective   Adverb

In this sentence, the adjective 'steroid' tells the reader what type of hormone testosterone is, and the adverb 'widely' indicates to the reader the range of activity within the body (as a rough guide, adverbs usually end in 'ly'), although you should note the comments already made regarding the use of relative terms.

If we wish to add still more information about testosterone, we could add a separate sentence:

> The steroid hormone, testosterone, is active widely in the body. It promotes the development of lean muscle mass.

In the second sentence, 'It' is the subject of the sentence, but refers back to the previous sentence describing testosterone. As we mentioned earlier, too many short sentences make the writing disjointed, so we can link these two sentences together:

> The steroid hormone, testosterone, is active widely in the body and promotes the development of lean muscle mass.

Here, the phrase 'promotes the development of lean muscle mass' is giving a more detailed description about the role of testosterone. Note that this is now a *phrase*, not a sentence,

---

BOX 6.1 Adjectives and adverbs

Adjectives are descriptive words used to describe a noun: 'a *brown* coat'. Adverbs are also descriptive and used to describe a verb: 'he ran *quickly*'.

because there is no subject within the phrase itself. The phrase is joined to the rest of the sentence by the *conjunction* 'and'. Conjunctions are words used to link phrases together, the most common being 'and' and 'but'.

With the aid of conjunctions, we can link many phrases together and end up with very long complex sentences, which become difficult for the reader to follow. On the other hand, as we have already mentioned, a series of very short sentences is also hard to read. So, how long should a sentence be? There is no specific ruling on this, but there are some rough guidelines you can follow. Most of the sentences in this section of the book are between 15 and 25 words in length. If your sentences are approaching 35 or 40 words in length (or more!) and comprise several phrases, you should think about dividing them up. Very brief sentences may be useful occasionally, for example, to emphasize a point. If, however, you find you have written a series of sentences, each of which is less than ten words in length, then see if the text will flow better if you combine some of them. The best test of your writing is to read it out loud. If you find it flows well and you can breathe in logical places, then you are probably about right.

6.3 Punctuation

Punctuation often causes students significant problems and written work may end up with commas, semi-colons, and other symbols scattered around seemingly at random. As we saw with the case of the unfortunate bar tender being shot by the panda, however, the presence or absence of a well-placed comma can completely change the sense of a sentence.

6.3.1 The full stop

The full stop is probably the easiest form of punctuation to deal with. Full stops are used to indicate the end of a sentence and are followed by a capital letter, which marks the start of the first word of the next sentence. Full stops are also traditionally used to indicate the use of a common abbreviation such as 'etc.' for *etcetera*, 'i.e.' for *id est* (meaning, that is), and 'e.g.' for *exempli gratia* (literally meaning, free example). In these cases, the full stop is not automatically followed by a capital letter. This usage of full stops for common abbreviations is decreasing and many writers simply put 'eg'.

6.3.2 The colon

The colon is most commonly used in the following ways:

- *as in the previous phrase, to introduce a list or an example*: 'A typical use of the colon would be as follows: . . .';
- *to introduce a quotation*: 'In his lecture, the professor stated that: "the nervous system . . ." ';
- *to link two contrasting statements that might otherwise be written as separate sentences*: 'The peripheral nervous system conveys information rapidly: this contrasts with the endocrine system . . .'. Here these two statements could have been written as separate sentences, but the use of the colon emphasizes the contrast between them;
- *to provide justification for a statement*: 'The peripheral nervous system conveys information rapidly: this is especially important when the body is responding to damaging stimuli'.

6.3.3 **The comma**

The comma is the most widespread form of punctuation and is frequently misused. Commas are used primarily as a means of providing a pause in the flow of the text; the pause can be in a variety of contexts, which are probably best illustrated by examples.

Separating items in a list

Commas can be used to separate items in a list: 'Oestrogen, progesterone, follicle-stimulating hormone, and oxytocin are all hormones that . . .'.

Separating out distinct phrases

Commas are used to separate out distinct phrases or clauses within a sentence. For example, we might write the sentence: 'Testosterone, which is one of the sex hormones, plays a role in the development of lean muscle mass.' In this context, the sentence would still be correct and have meaning if the clause 'which is one of the sex hormones' was not there. However, its inclusion adds further qualification to the statement.

The presence or absence of commas in the sentence can also change the meaning of that sentence. For example, compare the following two sentences:

> The adrenal hormones which are associated with the control of blood pressure, act on both the heart and blood vessels.
>
> The adrenal hormones, which are associated with the control of blood pressure, act on both the heart and blood vessels.

These two sentences have subtle, but important differences in meaning. In the first sentence, the implication is that, of the hormones released by the adrenal gland, only those that are associated with the control of blood pressure act on both the heart and blood vessels. By contrast, the second sentence implies that all the adrenal hormones are associated with the control of blood pressure, and act on both the heart and blood vessels.

Creating pauses

Commas are used to create a pause after the opening phrase of a sentence: 'In this essay, comparisons will be drawn between . . .' or 'As was stated earlier, the endocrine system is involved . . .'.

This is by no means an exhaustive list of the usage of the comma and we would encourage you to follow this up by reference to the more detailed expositions found in texts on English grammar, some examples of which are given in the 'Further reading'. As we have seen, the use of commas is important in imposing the correct structure and meaning to the sentences we write. To check whether you have placed your commas correctly, try reading aloud what you have written and ask yourself if the pauses indicated by the commas feel as though they are in the right place. This is by no means foolproof, but it will give you a useful guide.

6.3.4 **The semicolon**

The final common form of punctuation is the semicolon. Semicolons are employed much less frequently than commas. The semicolon can be thought of as being halfway between a comma and a full stop. It is typically used in the following ways:

- To separate items in a list where the sentence structure is complicated and commas are already in use to separate qualifying phrases, for example:

> The hormones that regulate blood pressure include: adrenalin, which is released by the adrenal glands and increases heart rate and peripheral resistance; angiotensin, which is released by the kidneys and . . .

- To link two related sentences together, rather than separate them with a full stop, for example:

> Adrenalin is released by the adrenal glands; it has to be transported in the bloodstream to the heart.

6.3.5 **The apostrophe**

The apostrophe needs special mention in any consideration of punctuation because it is probably the most abused of any piece of punctuation, although it is actually fairly straightforward to use. Increasingly, the apostrophe can be seen randomly used before or after the letter 's' when the 's' is simply indicating a plural. This is incorrect. Apostrophes have two main functions: to indicate possession and to indicate contraction.

Possession

To indicate possession the apostrophe is added to the word followed by the letter 's', thus: 'The heart's blood supply is . . .'. The test here is whether the sentence could be rephrased using 'of': 'The blood supply of the heart . . .'. If the word is a plural, already ending in 's', for example 'The kidneys', then the apostrophe is placed after the 's': 'The kidneys' blood supply is . . .' (do the test: 'The blood supply of the kidneys is . . .'). If the word is singular, but already ends in an 's', as is seen with some names such as 'James' then you still obey the rule for the singular form and add both an apostrophe and the 's', thus: 'James's heart condition worsened . . .'.

Contraction

An apostrophe can be used to indicate a contraction where some letters have been omitted because two words have been joined together to make a shortened form, such as 'don't' for 'do not' (remembering, of course, the earlier point in this chapter that you should normally avoid such contractions in academic writing!)

There are, however, some traps for the unwary:

> it's who's

These both indicate contractions: 'it is' and 'who is', e.g. 'It's clear that the experiment . . .'

Its whose

These are both possessive pronouns: 'The body regulates its blood pressure . . .'. Other possessive pronouns that don't take the apostrophe are: his, hers, ours, and yours.

6.4 Paragraphs

Solid pages of unbroken text are very off-putting and are difficult to read, so the text is usually broken up by separating it into paragraphs. A paragraph is usually made up of several sentences that are linked together by a common theme. The break between paragraphs can then be used to separate out different sets of arguments to help structure the logical flow of the essay as a whole. For example, if you want to make four main points in your essay, then give each point its own paragraph. A possible paragraph plan might look like this:

- introduce the main idea (topic sentence);
- explain the idea (amplify the topic sentence);
- present supporting evidence or examples (quotation, study, expert opinion, or report);
- comment on the evidence (show how it relates to the main idea);
- conclude the main idea (link to the title or link to the next point).

A brief sample paragraph from our essay on the peripheral nervous and endocrine systems could be structured as follows:

Introductory sentence, setting out the theme of the paragraph

Evidence to support the statement

Final sentence linking back to the introductory sentence

The speeds of communication of information are very different for the peripheral nervous system (PNS) and the endocrine system. In the case of the PNS, the conduction velocity of the action potential along the individual nerve axons can range from speeds of 0.5 ms^{-1} up to 120 ms^{-1}. The differences in speed depend on the axonal diameter and presence or absence of myelin, with the highest speeds being generated by the mechanism of saltatory conduction of the impulses along large-diameter, myelinated axons. The transport of hormones takes place through the vascular circulation. Here, the speed of flow also varies, but in the arteries it ranges from 0.02 to 0.2 ms^{-1}. The differences in communication speed between the PNS and the endocrine system are important considerations when comparing the functional roles of the two systems.

There is no rule that prescribes how many sentences make up a paragraph, but the point where you are moving from one topic to another within the essay makes a logical place to insert a break and start a new paragraph.

6.5 Abbreviations

Abbreviations are often used in scientific writing, usually to remove the effort from writing out lengthy terms in full each time they are used. The usual rule of thumb is that you should write out the term in full, followed by the abbreviation in brackets the first time you use it; on subsequent occasions you can just use the abbreviation:

> The electrocardiogram (ECG) was recorded while the patient.... The ECG recordings showed that...

In scientific convention, some abbreviations such as 'ATP' or 'DNA' are so commonplace that they are often used without previously being written out in full. However, the safest practice is always to write the term out in full on first mention.

Try this: Try out your punctuation

Now that you have read through these brief sections on punctuation, try using the rules to punctuate the following section of text. (Before you criticize us for not including any references in this piece, the references have been deliberately left out to simplify the exercise.)

The peripheral nervous system comprises neurons whose cell bodies are located in the ventral horn of the spinal cord for the motor neurons and the dorsal root ganglia for the sensory neurons the axons of the motor neurons leave the cord via the ventral root and join the axons of the sensory neurons to form a mixed nerve these nerves branch along their route from the spinal cord to the tissues and end by innervating the peripheral structures such as the muscles and skin some sensory axons terminate as free endings within the muscles and skin whereas others innervate specialized sensory end organs in the skin these organs include the Merkel disks Pacinian corpuscles and Ruffini endings in the skeletal muscles the large diameter sensory axons innervate specialized proprioceptors called muscle spindles and tendon organs the axons of the motor neurons innervate the muscle fibres of the skeletal muscles with each motor axon innervating several muscle fibres but each muscle fibre only being innervated by a single motor axon the axons terminate in a specialized structure termed the neuromuscular junction action potentials are conducted along the axon until they reach the pre-synaptic end of the axon here the action potential causes the release of the neurotransmitter acetylcholine (ACh) that diffuses across the synaptic cleft and binds with receptors on the post-synaptic membrane the binding of the neurotransmitter causes the muscle fibre to depolarize and fire its own action potential which results in muscle contraction.

A suggested model for drafting this piece can be found at the end of the chapter, but try your own version first.

6.6 Illustrations

You should always think about using illustrations in your essays and other pieces of course-work, since they can often be used to help you explain a process or describe a structure much more effectively than by using words only. Use of illustrations might include:

- flow charts to indicate the operation of a control system as shown in Figure 6.1;
- diagrams to help explain a life cycle or a biochemical process;
- drawings to help describe anatomical structures or the features of an organism;
- drawings of apparatus described in practical reports.

There are some key points to remember when including illustrations.

- Produce your own illustrations, don't copy-and-paste a diagram from a website: the rules of plagiarism apply as much to drawings as to the written word (see Chapter 5, 'Academic integrity and avoiding plagiarism'). Also, just copying a diagram will not help you to understand it.
- Give each illustration a figure number and refer to it in the text, don't just leave it as an 'add on'.

Figure number

Figure 6.1 Flow chart illustrating the feedback control of the release of growth hormone by the pituitary.

Explanatory figuretitle

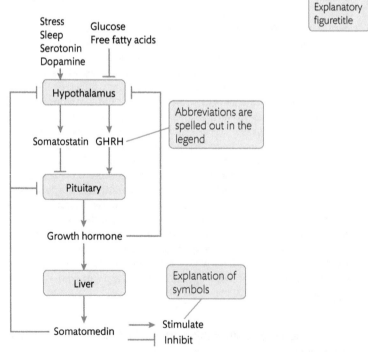

Abbreviations are spelled out in the legend

Explanation of symbols

GHRH, growth hormone releasing hormone. (Adapted from Berne and Levy (2000). *Principles of Physiology*. Maryland Heights, MO: Mosby, Figures 44–8.)

Citation for thesource of the figure

- Give each illustration a title so that the reader can see what is being shown.
- Make sure that your drawings are labelled so that the reader can identify the different structures being shown.
- Feel free to use digital software or freehand for producing your drawings: don't forget that in examinations you will probably have to draw your diagrams by hand.

Remember also that illustrations can save you words, so if you are struggling to write your essay within the word count, then a well-chosen illustration can save a lot of words in description or explanation.

Try this: Plurals and singular terms (see the end of the chapter for the correct answers)

What are the plural forms of the following words?

Nucleus, index, axis, analysis.

What are the singular forms of the following terms?

Data, criteria.

Try this: Style and punctuation

Rewrite the following sentences, correcting the punctuation and stylistic errors:

In our experiment it wasn't clear where the neurons' were located.
The data was not showing the distributions we expect.
The authors research didn't contains relevant data.

6.7 Referencing

Finally, in this chapter on choosing the right writing style, we need to flag up referencing. We dealt with this topic extensively in Chapter 4, 'Working with different information sources', and touched on it again in Chapter 5, 'Academic integrity and avoiding plagiarism'; however, it deserves a brief mention here because referencing is an important part of academic writing style. You must ensure that any information you get from another source is appropriately referenced.

 Chapter summary

Writing for a scientific audience requires a very formal style, with precise use of language. In this chapter we have looked at the conventions of writing style, such as the use of the passive form and an impersonal, rather than personal, expression. We have also revised some of the principles of basic grammar and construction. Finally, we have highlighted the importance of referencing, more of which can be found in Chapter 4.

Further reading

Seely, J. 2007. *The Oxford A–Z of Grammar and Punctuation*. Oxford: Oxford University Press.
Truss, L. 2007. *Eats, Shoots and Leaves: The Zero Tolerance Approach to Punctuation*. London: Profile.

Writing style exercise: suggested structure

The peripheral nervous system comprises neurons whose cell bodies are located in the ventral horn of the spinal cord for the motor neurons, and the dorsal root ganglia for the sensory neurons. The axons of the motor neurons leave the cord via the ventral root and join the axons of the sensory neurons to form a mixed nerve. These nerves branch along their route from the spinal cord to the tissues and end by innervating the peripheral structures, such as the muscles and skin. Some sensory axons terminate as free endings within the muscles and skin, whereas others innervate specialized sensory end organs in the skin; these organs include the Merkel disks, Pacinian corpuscles, and Ruffini endings. In the skeletal muscles, the large-diameter sensory axons innervate specialized proprioceptors, called muscle spindles and tendon organs.

The axons of the motor neurons innervate the muscle fibres of the skeletal muscles, with each motor axon innervating several muscle fibres, but each muscle fibre only being innervated by a single motor axon. The axons terminate in a specialized structure, termed the neuromuscular junction. Action potentials are conducted along the axon until they reach the pre-synaptic end of the axon; here, the action potential causes the release of the neurotransmitter acetylcholine (ACh) that diffuses across the synaptic cleft and binds with receptors on the postsynaptic membrane. The binding of the neurotransmitter causes the muscle fibre to depolarize and fire its own action potential, which results in muscle contraction.

Try this: Plurals and singular terms

What are the plural forms of the following words?

Nucleus—nuclei; index—indexes; axis—axes; analysis—analyses.

What are the singular forms of the following terms?

Data—datum; criteria—criterion.

Try this: Style and punctuation

Rewrite the following sentences, correcting the punctuation and stylistic errors:

In our experiment it wasn't clear where the neurons' were located.
In the experiment it was not clear where the neurons were located.
The data was not showing the distributions we expect.
The data were not showing the distributions that were expected.
The authors research didn't contains relevant data.
The authors' research did not contain relevant data.

Writing essays

 Introduction

Essay-writing is a very common form of assessment in almost all degree courses in the biosciences, both as coursework and for examinations. For many students, however, writing essays at university represents a real challenge, as it is a skill that is only developed to a limited extent in pre-university courses. It may also be a long time (perhaps years) since you last wrote an essay, and that might have been for an English course, rather than a piece of scientific writing. This chapter is designed to help you appreciate what is expected of you, and guide you through the process of planning and writing essays. Before we begin though, consider this point: we often refer to the process of producing an essay as 'essay-writing', but how much of the time is actually spent writing?

Think back to the last coursework essays you produced: how much of the total production time was spent writing and what else was involved in the process? If you are like us, then your time allocation for essays is probably something like this:

- quite a few hours searching for information in textbooks and research papers, and even more time reading through the material trying to understand it;

- following that, deciding and noting what is specifically relevant and should be included;

- then some sort of planning stage, where you decide how to order the material, and how to begin and round off the essay;

- finally, the writing itself, which probably only represents about 20% of the total time spent working on the essay.

7.1 What is an essay?

So exactly what is an essay? Dictionaries and websites abound with definitions of essays, viewed from different perspectives, some being more helpful than others. In historical terms, the word 'essay' means an attempt or trial. The *Oxford English Dictionary* includes the following definitions:

> The action or process of trying or testing.
> A composition of moderate length on any particular subject, or branch of a subject; originally implying want of finish, 'an irregular undigested piece' (J.), but now said of a composition more or less elaborate in style, though limited in range.

YourDictionary.com defines an essay as follows:

> A short literary composition on a single subject, usually presenting the personal view of the author.

One of the problems is that the definition depends on the subject area and the expectations of the readers. Perhaps the easiest approach is to consider what we might call the 'typical' essay in the biosciences. This would be a piece of structured writing, of 1000–3000 words in length, that is written as an answer to a specific question or in response to a specific title. In this context, the student is expected to be able to display knowledge and understanding of the subject material, setting it out in a coherent manner and drawing logical conclusions from the material presented in order to answer the original question.

7.1.1 What are essays supposed to achieve?

This is probably a question many students have asked themselves, particularly during the early hours of the morning as a deadline is getting very close! It may also be a question that markers have asked themselves, too, as they wade through a pile of exam scripts or coursework essays. Actually, a well-designed and well-answered essay question involves the development and testing of a significant number of skills, some of which are not explicitly expressed. It is all too easy to focus on the scientific content of the piece of writing and not to think much about the way that content is presented and structured, but the skill of writing is a critical element in the production of a good essay.

If we take the brief for a piece of coursework such as:

> In 1500 words, compare the endocrine and peripheral nervous systems as channels for control and communication within the body. The essay is to be submitted to Dr Harrison by 12 noon on Friday 1 March.

we can identify a list of the skills being exercised in order to complete the task as follows:

- using feedback from previously marked essays to produce a better piece of work;
- analysing the question;
- researching the subject matter;
- bringing together the information gathered from a range of sources, e.g. lecture notes, textbooks, research papers, web articles, etc.;
- planning the presentation of the information to address the question and make sure that all the key points are covered;
- ordering the material in a logical manner;
- framing the body of the text with an introduction, to set the scene, and a conclusion, to summarize the points made;
- writing succinctly and with clarity, paying attention to the grammar, syntax, and spelling;

Figure 7.1 Your first essay (probably!)

> It was a lovely hot day and I went to the beach with my sister. We fought a lot and Mummy got cross.

- referencing the information correctly to acknowledge the original sources;
- proofreading the essay to pick up on editorial mistakes and ensure clarity of writing;
- managing your time, so you are not scrabbling around trying to write the essay an hour before it is due in, even though you were given the title four weeks previously.

This is a surprisingly long list and it probably could be extended. The important point is that, while some of these skills, such as researching the topic, are obvious and easily recognized as part of the process, others, such as using feedback and proofreading, are more implicit, but are still very important if you are to write a good essay. As we go through the whole process of preparing and writing the essay, we will consider these different skills. Other aspects are covered elsewhere; using feedback in Chapter 2, researching the material in Chapter 4, and writing style in Chapter 6.

7.1.2 Principles of scientific writing

Before we can get to grips with answering the essay question, comparing the endocrine and peripheral nervous systems, we need to consider some of the features of scientific writing. Most people can remember the very first essays they wrote at school, probably on the theme of 'What I did in my holidays' or something similar. These essays are usually written in an informal style and referenced very much to the individual (Figure 7.1).

Writing for a scientific audience, whether it is an essay, a report, or a research paper, is a very different exercise and requires a different, more formal style, as we explained in Chapter 6, 'Choosing the right writing style'.

7.2 Approaches to essay-writing

Producing a coursework essay involves a number of key stages. In order to improve your essay-writing it is vital that you refer to feedback on previous work; this is something that you should bear in mind throughout the various stages, which include:

- analysing the question/title;
- researching the topic;
- making a plan;
- writing a first draft;
- reviewing and redrafting;
- proofreading, and checking citations and references.

Figure 7.2 Ideal allocation of time for essay-writing tasks

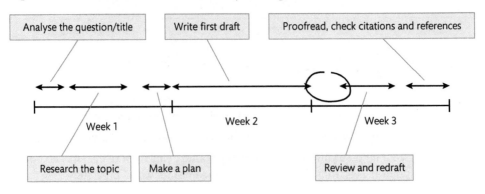

Ideally, these stages would be allocated appropriate portions of time to ensure that the production of the essay is as smooth and straightforward as possible, as represented by the timeline in Figure 7.2. The division of time won't be precise and doesn't need to be the same for each essay, but you do need to allocate time for all the important elements that make up the production of an essay, not just, for example, the writing. Importantly, the timeline also includes a gap between writing a first draft and writing a second draft. This will ensure that you can be more objective in your review and make significant improvements to the work.

7.2.1 Common problems

Figure 7.2 represents an ideal approach and we would encourage you to aim to allocate time to your essay-writing tasks in a similar fashion. If you can allocate your time in this way, it is a reliable method for increasing your chances of completing an essay on time and to an appropriate standard. However, in practice, the process of producing an essay is rarely as smooth and straightforward as this. Some common problems faced by students (and academics) when writing are as follows:

- leaving things to the last minute;
- doing too much research;
- doing little or no planning;
- taking too much time over the writing.

All of these are real problems, many of which you may have experienced for yourself. Each of the problems has a range of possible causes, and what causes the problem will determine what you need to do to address it. We will deal with each in turn.

At the last minute

You might find yourself trying to complete your essay at the last minute because:

- you have lots of other deadlines to meet;

- you feel anxious about writing essays and so avoid starting them;
- you just struggle to summon up the motivation to start earlier.

If you write an essay like this it means that you can be very focused and the work is often completed in the minimum time (because it has to be!) However, it can be a very stressful way to write and will affect the quality of your work, because you run out of time to review it and so end up handing in what is effectively only a first draft. If you find yourself in this situation, your immediate priority is to organize your time more effectively (as shown in Figure 7.2) so that you can allocate time to improving your essay by reviewing and redrafting.

Too much research

You might find yourself undertaking too much research because:

- you get side-tracked by information that is interesting but not relevant;
- you do a lot of reading which generates a lot of notes;
- you make very detailed notes on what you read.

If you write an essay like this it means that you have lots of material for your essay and you feel well informed; however, the volume of information can be overwhelming, not all of it is relevant, and there is less time available for other stages of the essay-writing process. Researching the topic is a vital task—nevertheless, you will not be judged on the information you have gathered, but on what you have written down. If you find yourself in this situation, your immediate priority is to develop a clear focus for your reading and note-making (see Chapter 4, 'Working with different information sources'). This will make you more efficient in gathering information, so you can then allocate sufficient time to the other stages. You may also be spending too long on research because you are not very effective at finding the material you need. Again, have a look at Chapter 4 for some tips about search strategies.

Little or no planning

You might find yourself doing little or no planning because:

- you run out of time (see 'At the last minute');
- you find planning a bit restrictive and you would rather just start writing, because writing helps you to think.

If you write an essay like this, because the writing process helps you to think, it may mean that your writing will flow well and so it will read quite fluently. However, it can be a confusing and inefficient way to write, because it is difficult to get an overview of the material and the logical sequence of the material can suffer as a result. If you find yourself in this situation, your immediate priority is to set aside time for planning. The aim of planning is to provide you with an overview of the essay content so you can select and order your points in the most effective way. The essay plan can be in linear note form or it may be in the form of a visual plan (see

Section 7.3.4, 'Make a plan'), but whatever the format, it will provide you with instructions for selecting and ordering the content of your essay.

Writing takes too long

You might find yourself taking too long over the writing because:

- you rewrite individual paragraphs again and again;
- you write too much, then struggle to cut it down to the word count specified;
- you simply find getting the words down on paper very difficult.

If you do write an essay like this it means that the writing gets a lot of attention, which is a good thing. However, it can be an inefficient and frustrating way to write and won't leave you with enough time for other aspects of the process. If you find yourself in this situation, your immediate priority is to speed up the production of the first draft, in order to have more time for reviewing and redrafting.

In this case, you need to have a plan of action for the writing process. Start with a set of bullet points that summarize what you want to say in the essay, and then just write a paragraph about each point, allowing yourself an approximate number of words for each paragraph as a proportion of the essay as a whole. Don't try rewriting any sections until all the paragraphs are written. When you have the body of the essay in place, you can think about linking the paragraphs together. At this stage, you will have a good appreciation of how long the essay is going to be and also of its overall structure. Now you can devote some time to editing or restructuring the essay.

7.3 Producing a coursework essay

We can now use the steps identified in Section 7.2, 'Approaches to essay-writing' to show you how to produce a coursework essay, based on the following stages of production:

- plan your use of time;
- analyse the question or title;
- research the topic;
- make a plan;
- write the first draft;
- review and redraft;
- use feedback effectively.

7.3.1 Plan your use of time

You will have noticed from the common problems identified previously that crucial to improving your writing is managing your time. Many of us (the authors included, as we are sure the Editor would confirm!) are not very good at time management. When we are set an assignment and told that the deadline for submission is not for six weeks, the natural reaction is often to put it in a drawer and forget about it until a few days before the submission date. That means that many of us err towards being 'at the last minute-ers'. We all know that this can lead to problems.

- 'I've only two days in which to write the essay and have just found out I can't get hold of a copy of the textbook because other students have already borrowed it.'
- 'The library doesn't have access to the journal article I really need and I have to find something else. Help!'
- 'The essay for this module is due in at the end of the week, but I've also got two other deadlines coming up . . .'.

So when should you begin working on an essay assignment? The obvious answer is 'as soon as you are given the title and your instructions'. It may not be practicable to race off and start writing immediately, but you do need to incorporate your assignments into your course planning. Clearly, the sooner you start the better, but you may need to consider some other factors as well. For example, although you have to submit the essay in week 8 of the term, you know that you won't finish covering the topic in lectures until the end of week 4, so you are probably better off not trying to do anything until you have the background from the lectures. An example is shown in Figure 7.3.

Figure 7.3 Planning your time for an essay

1	2	3	4	5	6	7
			Essay set	Analyse question		
8 Key lecture	9 Research topic	10	11	12	13	14
15 Planning	16	17 Writing	18	19	20	21
22	23 Reviewing & checking	24	25 Essay due!	26	27	28
29	30	31				

One useful approach to your planning is to put all your assignments onto a single chart or diary. For example, you could use a calendar. When planning, it is best to work backwards from the submission date and break the assignment down into its separate elements, as described earlier. This also has the advantage of giving you positive feedback because you can reward yourself as you complete each section on the calendar. If you make sure that you incorporate the submission dates for other pieces of work as well, then you can plan your time effectively, and also avoid hitting an unexpected log-jam of work that needs doing and handing in all at the same time.

When planning your timeline, you need to be realistic about allowing yourself enough time for the different components of the work; for example, carrying out research can be very time-consuming, particularly if you need to search for and read through (and understand!) a number of research papers. By the same token, you also need to be strict with yourself when undertaking the research, and not get carried away looking for more and more papers.

Obviously, you won't be spending all your waking hours during those two weeks working on the essay: you will have other commitments, such as lectures and practical classes, as well as keeping time for socializing. Therefore, within your timeline, you will need to decide on what days you are going to focus on the essay and for how long. Again, the key features are that you should try to be realistic, and then do your best to stick to your schedule.

What happens if things go wrong? Advance planning is again the key: whenever possible, leave a gap between your planned completion of the work and the submission date, then you have some leeway if something unexpected throws your plans out of line.

7.3.2 Analyse the question or title

It is very important that you write the essay that is asked of you: there is no point in writing an essay that doesn't address the question asked, however brilliant that essay may be. So, you have your time plan in front of you, and are ready to go. STOP. Before you do anything else, read the question carefully and then read it again, highlighting the following features:

- the subject of the question or title;
- the instruction;
- the key aspect;
- other significant words.

Figure 7.4 Analysing the question or title: example 1

Figure 7.5 Analysing the question or title: example 2

Figures 7.4 and 7.5 show a couple of worked examples for you.

Sometimes it might seem that you are having to highlight almost every word, but it is still worth doing if it helps you register all the important pointers.

There are several standard instructions used for essay questions, some of which appear to be very similar, but with some differences in emphasis. The most common instructions are:

- *Compare*: this is asking you to identify and comment particularly on features in common, but also aspects of difference between two or more identified elements.
- *Contrast*: this is also a form of comparison, but focusing predominantly on the areas of difference. This may be included with compare, as in: 'Compare and contrast . . .', thereby emphasizing that both the similarities and the differences must be brought out with equal weighting.
- *Describe*: is asking you to write a detailed description of the subject.
- *Discuss*: is often used to mean the same as describe, but it can also be used in its more literal sense of discussing the merits or shortcomings of a particular statement or idea, as in the example given (Figure 7.5).
- *Explain*: this is often used in the context of making clear how or why something happens, for example: 'explain the process whereby the blink reflex habituates during repeated stimulation'.
- *Evaluate*: the strict interpretation implies some form of quantitative assessment of information, but this may commonly be used in the more qualitative context of evaluating an idea.
- *Give an account of*: this usually has the same meaning as describe.
- *Write an essay about*: again, this typically means describe.

Note that the example in Figure 7.5 includes the word 'and'. This was to ensure that both sections of the title were addressed: 'Discuss . . . and describe . . .'. Although the first section may be the more important of the two, you need to give sufficient time and effort to the second aspect, even if it does not necessarily amount to being half of the essay. The pitfall to avoid is writing a very long first section and then suddenly realizing that you still have the second aspect to deal with, so that it is tacked on, very clearly, as an afterthought.

Once you have identified what basic type of essay you need to write, check for some other key pointers that give you some specific details about the approach to be adopted. These may include the following:

- *Write an illustrated account* . . . this is emphasizing that you need to include diagrams or other forms of illustration;
- *Using named examples* . . . make sure that you use specific examples that illustrate your arguments;
- *With reference to the recent experimental evidence, discuss the process* . . .;
- *With reference to the research literature, compare the model for . . . with . . .*.

These groups of titles clearly show the emphasis that the question setter wants you to take in writing your essay. For example, it is not just a factual description that is required for the last two examples—the essay should be heavily weighted towards using the experimental evidence or recent research papers to support your arguments. As you progress through your studies, reference to research evidence should become an automatic part of essay-writing and may not be referred to specifically in the title.

Finally, make sure that you are very clear about the breadth of the subject material required in your essay: look for the key aspect that specifies a group or range. For example, in an essay on the endocrine system, the title might refer specifically to steroid hormones; a title about ecology might refer to a specific type of habitat; or a title about animal behaviour might refer to a specific group or groups of animals.

7.3.3 Research the topic

Researching the topic is probably the most time-consuming part of the whole exercise. Your lecturers will no doubt hope that you are really enthused by the research you undertake and that you will explore the subject in depth. That's great, but you need a good dose of pragmatism in your approach and to be aware that undertaking the research could be literally an endless task. Therefore, you need to keep your timeline in mind and, while working as efficiently as possible, be very clear that you will need to set specific limits to the amount of research you undertake. You can also save yourself a lot of time by some initial investment: sit down with a blank sheet of paper and brainstorm some ideas—simply write down, in any order, all the ideas you have—about what types of information you need for the essay.

You will also need a strategy for your research, particularly when using online journals and other web-resource search engines: it is very easy to end up with very large numbers of hits, many of which are not directly relevant to your needs. (See Chapter 4, 'Working with different information sources', for a discussion of approaches to searching.)

Use appropriate sources

There is an increasing amount of material available for research purposes, and you should be strategic and critical in the way in which you use these different resources. These resources are dealt with in detail in Chapter 4, 'Working with different information sources', but a summary is also given here. The main sources of information are listed and it is probably a good strategy to work through them in this order, so as to keep a clear focus. This approach also means you are starting with relatively simple, well-defined information and progressively moving to the more complex, wider sources.

- *Lecture notes and handouts*: these will often provide the starting point for your investigations, by reading through what your lecturers have already told you about the topic.

- *Textbooks*: you can use textbooks to expand your knowledge and understanding of the principles of the topic. If you are in the very early stages of your university career, textbooks may provide the bulk of the information you need for the essay.

- *Reading lists*: for many modules you will be given lists of suggested reading that provide additional subject depth and breadth, to enable you to develop your understanding of topics covered in lecture courses. These lists will normally comprise journal articles and similar materials that have been selected carefully by your lecturers as being appropriate for the module you are studying. You may even be given some suggested titles as a starting point for your essay assignments.

- *Research journals*: as you progress through your course, there will normally be an increasing expectation that you will be reading and using research papers that you have found in research journals in order to develop the subject content of your essays. Through your university library you will probably have access to useful search engines, such as Ovid, Web of Science, and Medline, that you can use specifically for searching through the vast number of research journals that are available in electronic format. As we mentioned in Section 4.2.1, 'Academic journals', you will need to be careful in developing your search pattern. You can do this by selecting the appropriate keywords and seeing how many hits you obtain. Gradually refine your keyword selection until you have a limited number of hits that should be closely relevant to the subject matter of your essay. You will probably still have several tens of articles in your list and reading all those in full would, clearly, be very time-consuming. Reading the titles of the papers your search has come up with is therefore the next stage of selection. When you have drawn up a manageable shortlist of the titles that appear most relevant, then you can go to the next stage of reading the abstracts and, from your final list, select those papers which you need to read in full (having first checked that you can get access to the full articles!)

- *Web resources*: there is a vast and ever-increasing volume of material available on the web and it varies in quality from the excellent to the downright wrong (Section 4.2.2, 'Web search engines'). When you are looking through the journals of professional scientific bodies, you have the confidence of knowing that the research papers have been reviewed by scientists who know the subject area very well. Therefore, you can be confident about the validity of the material being presented. By contrast, the material that is posted on websites may have been posted by experts in the field, but it may also have been put there by people who have little or no real knowledge, but who have views about a subject that they want to broadcast to the world. So, you need to be cautious and critical, and look for specific signposts that can indicate the reliability of the information presented: for example, if the website is published by a university or a professional body, then the information is likely to be accurate and reliable. If it is published by a commercial organization, for example, a drug company, it is likely to be accurate, too, but may also be subject to a certain amount of marketing 'spin', for example, to encourage take-up of a specific drug. At the other end of the spectrum, if the material is published by an individual or organization with no obvious professional qualifications or academic affiliations, then you should treat the site with considerable caution. Mention here must again be made of Wikipedia. As we noted in Section 4.2.2, 'Web search engines', Wikipedia is an online encyclopaedia to which anyone can contribute. It is continually increasing in volume, but because of its nature, the quality of the material is variable: much is of high quality and has been written by experts in their respective fields; some is less accurate and is not necessarily reliable; some is wrong. Unless you are an expert who already knows the field, the problem you will have is that of picking out what is good from what is not.

Make relevant notes

You now have in front of you a large amount of information that you have gathered from your research. Most of it will be in the form of articles extracted from different sources. Some of these articles may be quite lengthy and all of them will contain more material than is needed for your essay. So the next stage is to extract the relevant sections from these articles and put them in note form in your own words. Section 4.3.1, 'Generic note-making, Notes are notes!' should guide you here.

7.3.4 Make a plan

There is often a great temptation, having done your background reading, simply to sit down and start writing. This is, generally speaking, not a good idea: it is very easy to drift off the topic of the essay, to leave important points out, or to produce a very disjointed essay. Planning is a very important part of essay-writing and will help you gain the best possible marks by allowing you to produce a coherent structure to your essay.

The exact timing of when you do the planning, however, will depend on your approach; there is more than one way of using plans in essay-writing. An essay plan can be used before or after the writing of the first draft, and your preference will depend on how you approach the task of writing essays.

As discussed earlier in this chapter, conventionally, the order would be:

- analyse the question or title;
- research the topic;
- make a plan;
- write a first draft;
- review and redraft;
- proofread, and check citations and references.

Not forgetting throughout the process to take note of previous feedback.

However, if you are someone who struggles to write an essay plan, what should you do? The answer to this question depends on why you struggle to plan. Consider the following descriptions and think which approach best describes your way of working.

Writer 1

'I like to make plenty of notes before starting. I like to have a good idea of what I am going to write before I begin the first draft. I find it difficult to begin a draft if I am not clear on the direction my writing is going to take.'

If this is you, then you need to use essay plans before you begin the draft, as per the conventional approach.

Writer 2

'I tend to make fewer notes than most people and tend to want to start writing as soon as possible. Often I don't know what I want to say before I start to write, but I find that ideas occur once I have started. For me, most of my thinking is done through the process of writing.'

If this is you, then you need to set an early deadline for the completion of the first draft so you have plenty of time to review and reorganize the content. This is important because the content of your first draft will only be organized in the order in which the ideas occurred during writing. You will therefore need to spend more time on reviewing the draft than a person who has planned the structure beforehand. Your approach to writing can be very effective, but you must allow plenty of time to review and improve the structure of the first draft.

Writer 3

'I like to make plenty of notes beforehand and probably have a general idea of the essay content before I begin to write. Once I have started to write I find that new ideas occur to me so the content of my draft is often different from my original intentions.'

If this is you, then you need to use essay plans before you begin the draft, as per the conventional approach.

Let's go back to one of the essay titles we discussed earlier:

> Compare the endocrine and peripheral nervous systems as channels for control and communication within the body.

There are many different ways of planning your essays, so we will look at just two different formats: the linear plan and the visual plan.

Linear plan

An example of a linear plan is shown in Figure 7.6. This is very useful if you like producing lists of key headings, to which you can then add sub-headings and specific details. Having drawn up the list of key points, you can easily reorder them to create a logical structure for the essay.

Figure 7.6 An example of a linear plan for writing a course essay

Introduction (approx 200 words)

Overviews of the endocrine and peripheral nervous systems
Descriptions of control – making things happen – and communication – sending
information

Body of the essay (approx 1600 words)

- Mechanism of communication (200 words)
  - Endocrine – hormones transported via blood stream, slow, widespread action –
  e.g. thyroid hormone
  - PNS – action potentials conducted via axons, fast, specific to a locality
- Action at cellular level (300 words)
  - Endocrine – can be via specific receptors on cell surface triggering second
  messengers – e.g. adrenalin
  Or crosses the cell membrane to act directly on the nucleus, e.g. testosterone
  - PNS – acts via specific receptors on the cell membrane – causes a change in the
  electrical potential of the membrane – e.g. acetylcholine
      - Point 3 (300 words)
        - . . .
      - Point 4
        - . . .

You can also see very quickly, by referring to your notes, if you have any gaps in your research that need some additional detail before you start writing. Furthermore, once you know how many key points you have, you can divide up your word allocation between the different points, with the target of giving a paragraph over to each point. When you do start writing, you can tick the points off as you go through them, and so make sure that you don't leave anything out. This approach also means you are not likely to drift away from the topic, which can happen if you just start writing without any planned sequence to the essay.

Visual plan

An example of a visual plan for the same essay is illustrated in Figure 7.7. Here, the plan takes the form of linked ideas that are gradually developed into more specific details. This is often a very useful approach if the essay is conceptually difficult to put together and you are not sure how you can structure it. Once you have your basic plan drawn, you can allocate the numbers of words to the different levels so that each major branch is given appropriate weight. Again, as you write, you can tick off the branches that you travel along. For example, starting from the centre, you can tackle communication as your first topic. So you can write a paragraph comparing the routes by which the communication occurs, namely the bloodstream and the

Figure 7.7 Developing a visual plan for an essay

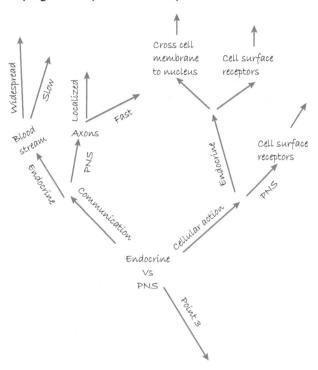

nerve axons; then moving to the outer branches, you can compare the characteristics of each means of communication. When you have completed this topic, you return to the centre point and start along the next main branch.

Try this—Produce an essay plan

Imagine that you are going to write an essay on your favourite hobby or interest (so that you don't have to do any research). Produce a plan for your essay. Get someone to give you some feedback on it. Does it look logical and well structured?

7.3.5 **Write the first draft**

There are three main sections to an essay that you will need to write:

- the introduction;
- the body of the essay;
- the conclusion.

We will look at writing each of these sections later in the chapter, but it is useful to have in mind a structure for the essay before we begin trying to write. One way of summarizing what these sections comprise is: 'Tell your readers what you're going to tell them, tell them (in detail), then tell them what you've told them.'

The introduction to the essay is used to frame the question and to set out what aspects are going to be addressed. An introduction is typically one or two paragraphs in length and should certainly not comprise more than 10% of the essay word count. The body of the essay is the bulk of the essay. In this section you set out the arguments, in detail and in a logical order. Finally, the conclusion is used to draw the reader back to the main focus of the question and to summarize the key points that have been made in the body of the text in a way that clearly addresses the question. As with the introduction, this should not normally exceed 10% of the essay.

Body of the essay

Regardless of when and how you plan your essay, you should have gathered together from your research the evidence to support your points. The inclusion of this evidence is a very important part of scientific writing. It is not sufficient to state:

> The endocrine system has widespread actions on the cells and tissues of the body.

Each statement should be supported by a specific example, so:

> The endocrine system has widespread actions on the cells and tissues of the body. For example, the thyroid hormones act to stimulate the metabolic rate of all the body's organ systems (Berne and Levy, 2000).

Note that the statement of the example is supported by a citation, indicating where the idea you have described came from (see Section 4.5, 'Citations and references' for more information on referencing). As you write each section of the body of the essay, make sure that you mark off each point on the plan and then make sure that the paragraphs link together, rather than appearing as completely independent sets of statements.

If, for example, you want to make five main points in your essay, then give each point its own paragraph. A paragraph is usually made up of several sentences that are linked together by a common theme (Section 6.4, 'Paragraphs'). The break between paragraphs can then be used to separate out different sets of arguments to help structure the logical flow of the essay as a whole. A possible plan for each paragraph might look like this:

- introduce the main idea (topic sentence);
- explain the idea (amplify the topic sentence);
- present supporting evidence or examples (quotation, study, expert opinion, or report);
- comment on the evidence (show how it relates to the main idea);
- conclude the main idea (link to the title or link to the next point).

Introduction

The introduction is a signpost for your reader, showing how you intend to answer the question. Remember what we said about the structure of the essay—in the introduction you are going to tell the reader what the essay is going to cover. One possible structure for an introduction could be:

- begin with a general point about the central issue;
- use the words of the title to show how you will focus on the question;
- indicate what the structure of your writing will be;
- make a link to the first point.

Note that in our consideration of the writing process, we have placed the introduction after the main body of the text. There is no rule that states that you have to write the sections in the order in which they will appear in the final document. Some people find it is logical to write that way, but others prefer to write the introduction after the rest of the essay, because then they have a clear picture of what it is they are introducing.

Conclusion

The conclusion is where you summarize for the reader your main points, linking back to the question and highlighting the most important aspects of your material. It acts as another signpost to your reader, rounding off the essay. As such it gives you the opportunity to:

- refer back to the question to show that your essay directly addressed the question set;
- summarize briefly for the reader the main points your essay covered;
- show the overall significance of the material in relation to the question;
- provide an overall assessment of theories or arguments, summarizing your own viewpoint.

Again, you might want to write this in a different order from the final document: for example, you might like to have a conclusion written out first because you can use it as a guideline for reminding yourself where the essay is heading.

7.3.6 **Review and redraft**

Reviewing and redrafting are important stages of the whole production process, but they often get overlooked because not enough time has been left to do them effectively. Again, this is where planning your use of time is so important—you should aim to leave yourself time after writing the original draft to review what you have written, rather than simply aiming to get the basic writing done by the deadline. Indeed, review is most effectively done when you have left a gap of a few days between writing the draft and undertaking the review. As identified in Figure 7.2, if you read through the essay immediately after you have written it, it is often difficult to look at it critically.

There are three key stages to the process of review:

1. Reading through the essay critically, ask yourself:
    - Does the essay answer the question?
    - Is there a logical flow from one paragraph to the next?
    - Does the introduction set the context of the essay and explain what is being written about?
    - Does the conclusion summarize the key points and set out the conclusions you have drawn?
    - Have I put in citations to support the key statements in the text?
    - Are any diagrams or illustrations in the right place, and have I included the source reference if necessary?
    - Have I made use of feedback from previous essays?
2. Redraft the essay to improve the structure following your review.
3. Proofreading the essay to:
    - correct errors of spelling and grammar;
    - check that the citations match the reference list.

Redrafting

Assuming that you have written your first draft in good time, you may wish to redraft the essay. The extent to which writers may wish to write more than one draft of an essay depends on how careful the initial writing stage was (as illustrated with the examples of writers 1, 2, and 3). Some writers prefer to get something down on paper (more typically on screen) and then juggle around with it until they are happy with the structure; others work very closely to a structured plan and, therefore, may only write a single version of the essay.

Redrafting can take two stages:

- reordering of paragraphs to help improve the flow of the essay;
- then re-writing individual paragraphs to improve the way in which the points are conveyed to the reader.

If you have written the essay on screen, rather than on paper, then reordering is relatively straightforward by cutting and pasting the paragraphs into a different order. Likewise, with the individual paragraphs, it is easy to rewrite a paragraph and insert it back into the text.

Don't forget to re-read the whole essay, to make sure that your reordering of the paragraphs hasn't broken the links between the different points.

Proofreading

It is very tempting, when you have finished writing the last word of the essay, just to submit it and sit back feeling satisfied that you have finished and it is on time (maybe only just!) It is very important, however, that you proofread the essay, checking for errors in your typing, spelling, and grammar. There is a danger that writers become too reliant on the spelling and grammar checkers in word-processing packages and assume that because the program has not highlighted a piece of text, then it is correct. These packages are not infallible. For example, many typographical errors will in fact result in a valid word so the spellchecker may not highlight it (e.g. 'to' and 'too' are both valid words, but they clearly have different meanings). Conversely, many programs will flag up many of the technical terms you may use in the biosciences. For example, the word-processing program used for this book identified words such as 'angiotensin' and 'osteoblast' as being incorrectly spelt. As a consequence your essay on screen may be highlighted in lots of places and so there is a tendency to ignore all of these as 'false positives'. To help avoid this it can be helpful to add the technical words you use regularly to the spellchecker's dictionary.

How should you go about proofreading? One of the hardest parts of proofreading is that when we read through something we have written we read what we expect to see and can very easily overlook mistakes. This is especially true if you have just finished writing the essay and are still feeling totally immersed in the exercise. If you have managed your time carefully you will have at least a few days spare before you have to hand in the essay. So, put the essay away for a few days, then sit down and read it carefully, as if it had been written by someone else and you were tasked with finding errors.

If you look to the punctuation section in Chapter 6, 'Choosing the right writing style', the best way of checking this is to read the essay out loud, as if to an audience. You might feel really stupid doing it, but you can always shut yourself away so no one will know! As you read, make sure that there are breaks in the text (and that you don't run out of breath!), and that they come in the right places for the sense of the text. Reading out loud is also a good test to make sure that the structure as a whole makes sense.

Finally check your use of references:

- make sure that your key statements are supported by citations in the text;
- check that each citation in the text is matched against a reference in the listing at the end of the essay;
- likewise, check that each reference at the end appears as a citation in the text.

Is proofreading really that important? Yes, it is for a number of reasons:

- the person who is marking your essay will not be impressed by reading text that is littered with errors and, as a consequence, you will probably get lower marks than the scientific content alone might merit;
- some mark schemes include a specific allocation of marks for grammar and spelling;
- errors in punctuation and syntax can lead to ambiguities of meaning, so the sense of what you are writing is no longer clear;
- it is important to get into the habit of writing accurately: if you are asked by an employer to write a technical report, that employer will not be pleased if there are numerous mistakes;
- furthermore, when you are trying to sell yourself in job applications, these mistakes can be very costly: employers will often filter out applicants for poor spelling or grammar.

So, get into good habits early and always check what you have written.

7.3.7 Use feedback effectively

It is very tempting, when receiving a piece of marked work, to look at the mark or marks obtained, feel moderately satisfied (or disappointed), and then put the essay away and not look at it again. It may well be that, since you submitted the essay, you have moved onto another module and don't feel that any comments would be relevant to what you are doing now. If you do this, you cannot hope to improve the way you write because you are not paying attention to the guidance given as to where you went wrong or how you could have done better. Likewise, it is also important to know what things you are doing particularly well, so you can continue doing them.

Using feedback is a very important part of the learning process and we covered this in detail in Chapter 2, 'Using feedback'. Feedback on essays should come in two forms: there should be subject-specific feedback, for example, the identification of factual errors or omissions, and more generic feedback related to your writing style. The subject-specific aspects should help improve your understanding of the topic and may be particularly useful when you come to revise for examinations. The generic feedback should enable you to improve your essay-writing for the future, irrespective of the topic. This generic feedback might address aspects such as the use of references, the layout of the essay, or the use of illustrations.

So, before you file your essay away, take time to read through it carefully. Re-reading an essay after a few weeks is itself a good exercise because you will have a very different perspective compared with when you were immersed in writing it. Read through and think about the feedback you have been given. First of all make sure that you understand why the essay as a whole got such good or bad marks: was it focused on the question? Did you go into enough detail? Did the essay have a clear structure with each point being explained and then leading on to the next? If you don't understand where you went wrong, make an appointment to see your lecturer so you can talk through the aspects you don't understand.

A really good exercise is to keep a note (on paper or online) of the key points of feedback from your different pieces of work. Note down the generic feedback from the current essay and then look back over the previous pieces of work to check whether this is an issue that has arisen before. If it is a problem that has been flagged up more than once before, you might well consider seeking specific advice about how to tackle it. Having noted down the key points, write a list of a few action points to bear in mind for the next essay you write, such as:

- write a more detailed plan;
- write shorter paragraphs;
- check the referencing style and that all references cited in the text are listed in the references.

Then, when you come to write your next essay, refer to your list of action points and make sure that you improve on those aspects. In this way, you should be able to maximize the benefit from the feedback given and progressively improve your writing. Don't forget, though, if you can't read the feedback comments or don't understand them, don't just give up—go and ask your lecturer for help.

7.4 Writing essays in examinations

Most bioscience courses at university will require you to write essays under examination conditions at some stage. In this section we are focusing on some tips in relation to essays; further guidance relating to examinations in general is given in Chapter 14, 'Getting the most out of exams'.

So, you are sitting in the examination hall with the paper in front of you giving a list of titles. You don't have three weeks to think about, research, and write each essay: you now have only three hours and you must write three essays in that time. After the initial panic, try to be very methodical in your approach and follow these key points: five minutes of care at the beginning of the exam can save you a lot of heartache later.

Read the instructions on the paper through carefully and slowly, paying special heed to any that specify how many essays you should answer and whether you have to select titles from specific sections of the paper.

- Read the titles through carefully.
- Eliminate any titles about which you know nothing (hopefully not too many!)
- Select a set of titles on topics that you can answer, then narrow that down to the requisite number by checking the titles again, highlighting the key words, and then selecting those titles that you are confident you can address in full.
- *Plan your first essay*: again, time spent planning the essay and checking against the key words in the title will help make sure that you answer the question.
- As you write the essay refer back to your plan and tick off the key points as you address them.
- Don't forget to include illustrations if they are relevant.
- *Be strict with yourself about timing*: it is very easy to select the first essay title as one you know most about and get carried away writing it, so that you don't leave yourself enough time to do justice to the other essays. So don't allow yourself to spend a long time on the first essay; divide your time up equally.
- Leave enough time at the end for reviewing.

If you only answer two essays instead of three, or select your titles from the wrong section of the paper, or don't address the question, you cannot hope to get good marks!

A final tip for this section

For your coursework essays you will probably have had a long time to write them, you will have written them using a computer, and probably did not write them at one sitting. Examination essays are very different: at the moment, most examination essays are written longhand, using a pen. So, it's a very good idea to get some practice in writing under these conditions; this can form a valuable part of your revision. When you have finished revising a specific topic, look out some past examination papers and find a question on that topic. Then sit down and write an essay by hand, under examination conditions. This will not only be a good test of how much you know and understand about the topic, but it will also give you a feel for how much you can write in the time available. As a final guide, ask a lecturer to read through the essay and comment on it, so you can confirm that you are on the right track.

 Chapter summary

Essay-writing is an important skill and, as with all skills, it develops with practice. In this chapter we have shown the importance of:

- planning your use of time as soon as you have the essay title, so you don't end up rushing at the last minute;

- analysing the question or title carefully, so you make sure you write the essay your tutors want;

- researching the topic carefully, using the different sources available to you;

- planning the structure and order of the essay;

- reviewing, redrafting, and proofreading to make sure you have answered the question, that the structure is logical, that the citations and references are correct, and that there are no spelling or grammatical mistakes.

Writing practical and project reports

 Introduction

The knowledge base in all areas of the biosciences is underpinned by careful experimentation and observation. The ability to design experiments, to undertake them methodically, to observe and record the outcomes accurately, and to analyse and interpret the findings is vital to the development of a scientist. These skills, however, are not sufficient on their own: you also need to be able to communicate your findings to other people. If someone asks you for directions to find somewhere, knowing where to go is only part of the issue; you also need to be able to communicate that knowledge. In the scientific community this is done in a number of ways, the most common being presentations at scientific meetings and published reports.

If you are planning to follow a career in the biosciences, you will need to develop all of these skills. However, such skills are not just of value in a scientific career—there are many careers where the skills of accurate recording, analysis, interpretation, and presentation of information are highly valued by employers.

During your studies you are likely to undertake a range of practical work, including practical laboratory work, fieldwork, and independent project work. The aims of these types of work are to develop your skills as a scientist and also to deepen your knowledge and understanding of the subject. The aims of this chapter are to help you get the most out of your practical work and to help you prepare high-quality reports. Although we will discuss fundamental aspects of data presentation, this will be fairly limited since there are other guides that go into this in great detail. Likewise, we will not discuss the processes of experimental design or data analysis. Good guidance for these aspects can be found in one of the partner books in this series (Moody, P., Dine, D. and Trueman, L. (2016). *Research Methods for the Biosciences, 3rd edn.* Oxford: Oxford University Press).

This chapter is laid out so that you can read it straight through, or dip into sections for guidance on how to tackle specific aspects of your work. During your studies you will be expected to read scientific papers and it is always a good exercise not just to focus on the scientific content, but also to evaluate the way in which the science is presented. As you read on, it will be helpful to have some research papers to hand so you can look at the different styles of presentation.

Although the focus of the chapter is on laboratory-based practical classes and project work, much of the material is also directly relevant to fieldwork as well. To help create some context, the text uses as its examples an undergraduate investigation into the jumping ability of the locust. The chapter has been structured to follow the timeline of the stages of the practical class from initial preparation to completing the report, as shown in Figure 8.1.

Figure 8.1 Structure of the chapter

8.1 Before the practical class

The writing up of your practical work is the final stage of the process, and you can only hope to produce a good write-up if you have undertaken the work effectively in the first place. If you turn up to the practical class not knowing what you are supposed to be doing and not having checked that you have brought everything you need with you, then you are certainly in danger of working less efficiently in the class and are also likely to perform less well. Good preparation beforehand is, therefore, a key stage to success in all types of practical work.

Much of this section will seem to be simple common sense, but if you want to get the most out of the practical work you are going to do, it is essential that you prepare properly. This is something that many of us are not very good at—it is very easy just to arrive at a class and hope everything will be made clear as you go along. However, not only will effective preparation help you learn more and get better marks, but it can often pay off in terms of saving you time and effort when you are actually doing the work. There are few things more frustrating than having spent a long time setting up an experiment only to discover that you have set it up incorrectly or not recorded all the data you need, so you have to start all over again.

Good preparation is particularly important if you are going to be working away from your normal site, for example, in the field, when you cannot easily go back to fetch something you have forgotten. More specifically, good preparation means knowing:

- where you have to be and when;
- what items you need to bring with you;
- what you are going to be doing and why;
- some background to the investigation;
- how you are expected to present your findings.

For many practical exercises you will be given a schedule beforehand that will provide most of this information, but the following steps will also help.

8.1.1 Read the schedule

Reading the schedule carefully will enable you to check the time and location of the class, and what items, such as a lab coat, calculator, etc., you need to take to the class. A lined notebook for recording your results and any notes about the experiment is always useful and much better than a borrowed sheet of file paper. It is good practice to get into the habit of writing your experimental notes in a notebook. This will be particularly important for a research project, for which you will be expected to keep a lab book that you may need to submit along with your report. Alternatively, you can keep your notes in electronic format, but make sure that, if required, they are in a form that can be submitted alongside your overall report. Most scientific employers will require you to keep detailed notes so that, if necessary, results can be verified at a later date.

8.1.2 Read the details of the experiment

Reading the details of the experiment will give you a good idea what you will be doing. It may well be that the techniques and equipment will be new to you, but familiarizing yourself with the description of the procedures will help when it comes to doing the experiment. Highlight any key sections of the methods, for example if specific measurements need to be made at specific times. Figure 8.2 shows an example of how you might do this for an experiment. If you are going to be working in a group, getting together to plan how you are going to organize yourselves and divide up the workload will help you to be more focused and save time.

8.1.3 Read up on the background

Reading up on the background to the work will help you understand the biology involved and can put what you are doing into context. It will help to read through your notes from related lectures. If there is a reading list, try to read through at least some of the material beforehand and make notes of key points (see Section 4.3, 'Note-making strategies'). There is also the bonus that, if the papers or texts are not available online and need to be borrowed from the library, you won't have to wait because other people have borrowed them first!

Figure 8.2 Using highlights to identify key parts of the experimental protocol

Jumping distance: experimental measurements
Measurements must be made in a warm environment (if possible 28–35°C).

You should try to obtain records from at least ten adult locusts. Each locust should be weighed and the length of the femur of the hindleg measured.

Take care not to mix up your locusts during the recordings: use ink-marks on the hindlegs for identification. The wings of the locusts should be fixed together with a small strip of adhesive tape to prevent them from flying.

8.1.4 Check if there are instructions about presenting your results

Checking how you should present your results is important because it may vary from class to class. Sometimes you will simply be required to fill in tables or draw graphs on the schedule itself and answer some specific questions, but at other times you will need to write a full report similar to a scientific paper (you will certainly be required to do this for a project report). Again, knowing in advance what is expected of you will help you to be more efficient in your work and save you time in the long run.

8.2 During the practical class

If you have done some preparation for the practical class it is much more likely that the class itself will go well. There are two important things you need to do during the class to ensure you can write your report afterwards—make notes and record your results.

8.2.1 Make notes

Try to get into the habit of making notes during your practical classes. Many practical classes will begin with a briefing session addressing a variety of aspects of the work, such as safety issues and instructions, and demonstrations of the techniques to be used, as well as some background subject information. You may find you have to stand up to see and hear what is going on but, even so, do take notes because otherwise you will forget everything very quickly. There will be other times during the practical class when it will be useful to make notes as well, for example, if you need to make changes to the procedures during the course of the experiment or if you discuss some aspects of the experiment with your demonstrator or lecturer.

Before beginning the experiment, check through the sequence of tasks that you have to complete and, especially for timed measurements, make sure you have a clear list of what needs doing and when.

At the end of the class, you may well have another briefing session. Make the most of the information you are given because this could be very important for when you are writing up: make good notes—don't just rely on your memory, because you will forget things!

8.2.2 Record your results

Recording your results is clearly a critical part of undertaking any piece of practical work. Increasingly, you will find that you are recording experimental data directly into a spreadsheet or other program. However, whether you are storing the information in a spreadsheet or in your laboratory notebook, you still need to draw up tables for the results, making sure you identify what is being measured and what each column of the table represents.

During the experiment record your results carefully, clearly identifying numerical values and the units in which they were measured. Also make notes of any changes you made to the protocol, or to the way in which the results were measured. Again, care taken at this stage will save you a lot of time and frustration later on. You are also much less likely to make mistakes. Figure 8.3 illustrates two very different ways of recording the results from a simple practical class measuring the distance locusts can jump.

Figure 8.3 Two very different ways of recording your experimental data

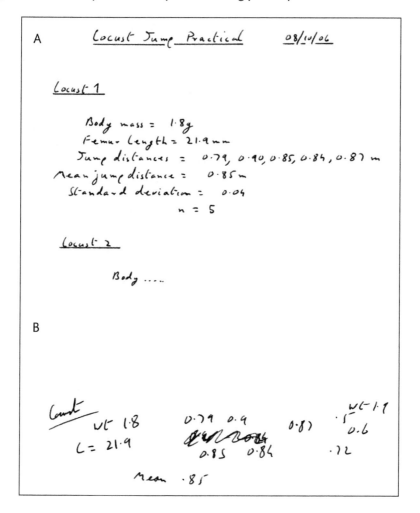

Compare the two sets of records in Figure 8.3 and list the aspects that you think are unclear in the data sets in sample B. If you were presented with these two data sets, would you be able to write about them in a meaningful way? Certainly, one advantage of recording your results directly into a computer is that you won't have to try to decipher a colleague's handwriting (or even your own!)

8.2.3 When things go wrong

What if things go wrong? By their very nature, experiments may go wrong, and you find you don't have any data or that they do not make sense. If this happens speak to your tutor as soon as possible—it is likely that you will be able to obtain another set of results from your tutor or from your colleagues in the class that you can then use for your report.

8.3 Writing your report

During your degree programme you will probably be asked to write up your practical work in a range of different formats. These may range from simply filling in tables and writing short answers to questions listed on the practical schedule you have been given, to preparing a full report in the form of a scientific paper. In this section we will take the scientific paper as the model, as it covers the most common aspects of report-writing. Make sure you are clear what format of report is expected from you before you write!

As we stated in the Section 5.1.3, 'Deliberate plagiarism', 'Collaboration and collusion', it is important to remember that, while you will often need to work with a fellow student to conduct a practical experiment or undertake fieldwork, this doesn't usually mean that you are being asked to collaborate on the writing-up aspect of the report as well. Obviously, there will be similarities between two reports written by two students who had worked together to conduct an experiment, but these will largely be confined to the methods and results sections. However, in the discussion section, the students will be expected to analyse and draw conclusions from their findings *independently* of each other.

8.3.1 Use an appropriate writing style

The principles of writing for a practical report are the same as those for other pieces of scientific writing and have been discussed in detail in Chapter 6, 'Choosing the right writing style'. The key point to remember is that you are writing for a professional readership and, therefore, you need to write in a formal, scientific style.

Avoid the use of 'I' or 'we'

So the phrase: 'I measured the locust jumps at ten-minute intervals . . .' becomes: 'The locust jumps were measured at ten-minute intervals . . .'. Likewise: 'On the basis of these findings, we concluded that . . .' becomes: 'On the basis of these findings, it was concluded that . . .'. This is known as writing in the *passive voice*.

Write in the past tense

You should describe what *was* done or what *was* observed. The only exceptions to this are when you are discussing illustrations or data presented in the report and the conclusions you can draw from them. For example: 'Figure 1 *shows* that jump distance *increased* with body mass . . .' or 'The key conclusion of this report *is* that jump distance *increases* with body mass . . .'.

Reference your sources

Wherever you take other people's ideas from, you must reference them correctly in the text and list the sources in the reference list at the end of the report. Remember that not to do so is plagiarism (Section 5.1, 'What is academic integrity?'). Referencing is dealt with in detail in Section 4.5, 'Citations and references'.

Use SI units

Measurements should normally be presented using the Système International d'Unités (SI) conventional form. The three fundamental units are the kilogram (kg), metre (m), and second (s). The SI system of units is based on seven 'base units', which are the fundamental measurements from which the other units are derived (Table 8.1). From these base units we derive a series of units for different measurements. Table 8.2 lists some of the most common units that you may come across in practical work and their definitions.

8.3.2 **Get the structure right**

The standard format of a scientific paper is as follows:

- Title
- Abstract
- Introduction
- Methods
- Results
- Discussion
- References.

You may also wish to include some additional sections, such as Acknowledgements, Appendices (see *Results*), and Keywords, or a list of abbreviations used.

Table 8.1 The seven base units of the Système Internationale d'Unités (SI units)

Quantity	Unit	Symbol
Length	metre	m
Mass	kilogram	kg
Time	second	s
Temperature	Kelvin	K
Amount	mole	mol
Current	Ampere	A
Luminosity	candela	cd

Table 8.2 Common units and their definitions

Measure	Unit	Symbol	Derivation
Area	square metre	m^2	
Volume	cubic metre	m^3	
Volume	litre	l	$0.001\ m^3$ $1\ dm^3$ (1 cubic decimetre)
Velocity	metre per second	$m{\cdot}s^{-1}$	
Acceleration	metre per second per second	$m{\cdot}s^{-2}$	
Force	Newton	N	$kg{\cdot}m{\cdot}s^{-2}$
Energy, work	Joule	J	$N{\cdot}m$ $kg{\cdot}m^2{\cdot}s^{-2}$
Power	Watt	W	$J{\cdot}s^{-1}$ $kg{\cdot}m^2{\cdot}s^{-3}$
Concentration	mole per cubic metre	$mol{\cdot}m^{-3}$	
Temperature	degrees Celsius	C	K
Electric potential difference	Volt	V	$W{\cdot}A^{-1}$
Electric resistance	Ohm	Ω	$V{\cdot}A^{-1}$

Note that when writing compound expressions, e.g. $m{\cdot}s^{-1}$, you should use the middle dot to separate the terms. This avoids confusion where terms could be misinterpreted; otherwise, for example, $N{\cdot}m$ could be mistaken for nm (nanometre) when hand-written.

We will look at each of these sections in turn and you may find it useful to refer to some of the papers from your reading lists to look at the way these sections are organized and presented.

Title

The title must be brief and informative, so that the reader can see at once what the paper is about. In the case of scientific papers for publication, the title is very important as it will be used as part of the information for electronic searching, so authors will often include key-words to increase the probability of the paper being picked up by an electronic search.

Often journals will limit the number of characters that can be used in the title. A typical limit would be 120 characters, including spaces, so it is good practice to write your titles within this limit.

Have a look at the four sample titles that follow, all of which describe the same investigation.

1. An investigation into the relationship between the body mass of an adult locust, of the species *Schistocerca gregaria*, and the maximum distance the animal can jump under non-restrained conditions.

2. The relationship between body mass and maximum jump distance for the adult locust, *Schistocerca gregaria*.

3. The locust jump.

4. Is bigger really better? Testing the jumping ability of different sizes of locust.

Which title provides the key information most efficiently?

The first title is unnecessarily long at 200 characters, although it does contain all the essential information. Here, the first section of the title 'An investigation into . . .' (or equally 'A study of . . .') is redundant and can be deleted without changing the sense of the title. The third title is too brief, since it does not give any indication of the aspect of the jump being investigated. The second title is 108 characters long and contains all the essential detail for the reader to know whether to read further or not. Slightly humorous titles, such as the fourth example, may well be appealing for certain reports, and are often styles used by the popular press, but this approach should be treated with caution!

Abstract

There is a real art to writing a good abstract. The abstract must provide the reader with all the essential information from the paper within a very limited number of words, typical abstracts being around 150–200 words in length. In many cases, the abstract may be used as a free-standing source of information: for example, when you are researching for an essay, you may well only read the abstract of a paper, rather than the whole paper.

The abstract will normally be the last part of the paper that you write and it is worth going through the following checklist to make sure you have covered everything.

- *Background*: very brief, but enough to set the context for the . . .
- *Aims*: the key question you set out to answer using the . . .
- *Methods*: again, these should be described very briefly, summarizing the technique(s) used to obtain the . . .
- *Results*: here, you need to quote the key findings of the investigation, including the actual values obtained from which you drew the . . .
- *Conclusions*: the key points you drew from the results. The conclusions should be related back to the initial aims.

The results and conclusions are the most important part of the abstract because, in a scientific paper, this is where you are presenting new findings. These two aspects should, therefore, make up the bulk of the abstract, with the background, aims, and methods being written as succinctly as possible. When you read through the abstract of your report (or better still ask a friend to read it through), the two points that should be immediately obvious are: what was found and why it is significant.

Compare the information content of the two sample extracts from abstracts describing our locust jump experiment and identify the key pieces of information that are missing from the second extract.

Abstract 1

The capacity of adult locusts to jump long distances to avoid predation is well known; however, it was unclear whether there is a relationship between the body mass of the adult locusts and their overall jumping performance. To investigate this, 20 male, adult *Schistocerca gregaria* (mean mass 1.6 ± 0.3 g) were stimulated to jump by means of a standard visual stimulus and the mean distance measured for ten jumps. A minimum of 15 minutes was allowed between each stimulus presentation to avoid habituation. The overall mean distance for the jumps was 0.95 ± 0.07 m and there was no correlation between body mass and mean jump distance ($r^2 = 0.15$). The findings indicate . . .

Abstract 2

Adult locusts can jump long distances. In these experiments the jumping ability of adult locusts was measured. The adults all jumped in response to a stimulus and the jump distances were averaged. Sufficient time was left between the jumps so that the response did not habituate. The results did not show any relationship between the mass of the locusts and the distance they jumped. The findings indicate . . .

Although the second abstract does describe the conclusion of the experiment, it does not give any indication of the initial aims. Nor does it provide any of the key information and numerical data to allow the reader to appreciate what was done and what was found.

As with the title, the abstract of a paper is commonly used as a means of electronic searching and so should contain keywords that identify the nature of the study and its conclusions.

It is not normal practice to include references within the abstract.

Try this—Write an abstract

Go to the Wikipedia pages for either animal navigation, plant hormones, or cell communication (or another biological topic that interests you) and write an informative abstract that would help future visitors to the page.

Introduction

The introduction is the section of the report in which you describe the background and aims of the study. You might like to think of this section as addressing four questions:

- What has been done before?
- What still needs to be resolved?
- Why is it important to resolve this question?
- What do I aim to do?

A useful approach to drafting the introduction is to start by setting out the aims of your study as a series of bullet points. Keep this list of aims in front of you while you are researching and writing the background to the study so that you keep clearly focused on the topic and don't start writing about sideline issues.

For a report in the form of a scientific paper you will need to read published papers on the topic of your investigation so that you can write a brief overview of the research that is underpinning your study. The skill in drafting this part of the introduction lies in writing a succinct summary of the findings of the previous authors and of the experimental evidence supporting those findings. As with all such writing, it is important that you present the material in your own words and fully reference the statements (see Section 4.5 'Citations and references').

The background text is normally presented as a factual account without discussion of the ideas presented in those papers. In most areas of research, there are areas of knowledge that are accepted as agreed and which you can describe as the background to the study. Developing from these, there will be areas where there is still disagreement, or questions remaining to be answered. In the case of disagreements in the published research, you will need to present both sides of the argument. This is best done simply by presenting your synopsis of the two arguments in sequence. Thus you might write:

> Derby and Joan (2010) concluded, on the basis of laboratory observations, that the locust jump is a behavioural response triggered by the presence of an attractive mate. However, in a more recent study undertaken in the wild, Bonnie and Clyde (2012) reported that the locusts only jumped to avoid being predated.

In the case of addressing questions still to be answered, you might phrase your statements as:

> Derby and Joan (2010) concluded, on the basis of laboratory observations, that the locust jump is a behavioural response triggered by the presence of an attractive mate. However, it is unclear as to whether this pattern of behaviour would occur in the natural environment.

Your description of the background to the study should, therefore, explain the scientific basis of the study, and provide an explanation of what is still not understood and why it is important that it should be understood. On that basis, you can then explain the aims of the study you have undertaken.

It is inevitable that, for most of your practical classes, you will not be aiming to resolve current controversies or discovering the solution to a specific question in science, although it is perfectly possible that you may be doing that in your final-year research project. For the most part, your practical work will be undertaking experiments that have well-known results. Despite this, it is important that you research the background to the study and present it as a synopsis, in your own words, of the scientific knowledge and can set out the aims of your experiment, based on that knowledge.

Methods

The methods section of the report is where you describe:

- what you did;
- how you did it;
- what you used to do it.

If your report is on a piece of fieldwork, you will also describe:

- where you did it;
- when you did it.

The methods section should be written in such a way that another scientist could repeat your work on the basis of your descriptions. Particularly when writing the methods section for your final-year project, you may need to include more detail than would normally be found in scientific papers, for example, regarding the composition of solutions used.

Think of this section in terms of preparing a recipe for baking a cake. When you read a recipe, you need to know:

- what ingredients to use;
- how much of each;
- how to mix them together;
- what type of baking tin to use;
- what oven temperature to bake it at and for how long;
- how to analyse the results (eat it!)

Although you are describing what you did, you should still remember to write in the passive voice: 'The locust jump was measured using...' rather than 'I measured the locust jump using...'.

Check how much you need to write

In terms of reporting on class practical exercises, the necessity for a methods section is variable. For example, where you have been given a detailed schedule to follow, you will quite probably be told that you do not need to copy out the methods again. Even in this case, however, you will be expected to report on any changes made to the protocols, particularly since these may affect the results you obtain.

Assuming that you will be writing a full report, you will need to describe the methods you used in detail. If you are using a new technique, or have been instructed to give full details of the protocols, then you should set them out in full. In most papers, however, if you are using a standard procedure that has already been described in detail in another paper, then you can simply refer to the previous paper and need only specify any changes you made:

> The locust leg muscles were stained for ATPase using the method of Peters (2004), but with an extended incubation period of 10 minutes.

Species names

If you are using animals or plants in your experiments, you should give the full name of the species, which is written in italics for the genus and species name, and the number of individuals involved. Where it is significant to the experiment, you should also give other details, such as the age, sex, and physical dimensions (e.g. height, weight). Thus, for the locust jumping experiment, your methods section might state:

> The jumping experiments were performed using 20 adult, male locusts (*Schistocerca gregaria*), mean weight 1.6 ± 0.3 g (standard deviation).

Human subjects

If you use human subjects in a research project, you need to have ethical approval from a local or NHS ethics committee for the work and you should quote the approval in the methods. You will also need the informed consent of the subjects, and this should also be stated.

Chemicals

When listing chemicals or other substances that were used you should give the name of the chemical in full (e.g. sodium bicarbonate) and/or its chemical abbreviation (e.g. $NaHCO_3$), along with the concentration of a solution and the amount used. If you have used specialist reagents you may also need to note additional details, such as the name of the supplier.

Specialist equipment

When describing the equipment used, again the amount of detail you need to give depends on the degree of specialization. For example, if you are stating that you used 3 g of sodium bicarbonate, you don't need to give details of the balance used to weigh out the chemical. However, you might have used a specialist piece of equipment for your experiment, for example, a confocal microscope or a specific computer interface for data logging. In this case, you should give the details of the equipment, such as the manufacturer and the model number. In some cases, it may also be helpful to include a simple diagram of the apparatus used, to show how different pieces of equipment were connected together.

Statistics

We will not go into any detail of statistical analyses as there are many books for the biosciences that address this topic. However, there are some key points that you must remember.

For many experiments you will be undertaking several measurements. You will therefore need to give descriptive statistics in your methods. The basic statistics would be measures such as the *mean*, the number of measurements taken (n), and the range of the values. For example, if you quote a mean value for a result, you need also to give the number of tests carried out (e.g. $n = 27$) and an indication of the spread of the data. The spread of the data is often given in the form of the *standard deviation*, which gives the reader a measure of the confidence that can be placed in the mean value. In the methods, you would state:

> Means are given with the standard deviation . . . the adult locusts used had a mean leg length of 21 ± 3.2 mm ($n = 20$).

You may also be employing statistical tests in your analysis of the data, for example to distinguish whether a treatment had an effect on the sample population. For most practicals, you are likely to employ routine tests, such as the Student's t-test, the Mann–Whitney U-test, or the chi-squared (χ^2) test. In such cases it is sufficient to state the name of the test and the level of significance that you are accepting, e.g. $P < 0.05$. For example, in the case of comparison of two groups of locusts you might state that:

> The mean leg length of the adult locusts was significantly greater than that of the juveniles ($t = 3.2$, $P < 0.05$, Student's t-test).

More extensive details regarding statistical analysis can be found in *Research Methods for the Biosciences*, 3rd edn (Holmes *et al.*, 2016).

After all that

When you have completed writing the methods section, read through it and check that you could repeat the experiments from the description you have written (think back to the recipe—could you bake the cake?)

Results

The results section is the core of the report. Here, you *describe* and *show* what was found along with the analyses of those findings. You should aim to lead the reader through the findings, highlighting the important features. You should not, however, attempt to interpret or explain the results in any way—this section is a factual description.

When writing the results section, you should use your data to tell a story. It is very important that the text you write can stand on its own in terms of communicating the key points. The graphs, tables, and other images are then used to support the description. A common error in students' results sections is for the data simply to be presented as the graphs and tables, with little or no explanatory text, leaving the reader with the task of trying to work out what is important. So rather than write:

> The results for the jumping experiment are shown in Figure 1.

it is much better to write . . .

> From Figure 1 it can be seen that there was no significant correlation between the distance jumped by the adult locusts and their body mass ($r^2 = 0.15$).

In the case of the first extract, it is left to the reader to draw their own conclusions from the plot shown in the figure. By comparison, the second extract provides the key summary of the observation, using the figure to illustrate the point and providing statistical confirmation, by means of the coefficient of correlation, that there was no relationship between the two variables.

Calculations and levels of accuracy

When using calculators and spreadsheets, be aware of the levels of accuracy of your measurements; often, a calculation performed on a calculator can result in a number with a string of up to ten digits. For example, when measuring the jump distance for our locusts we might only be able to measure the distance to the nearest centimetre. So an individual measurement might be recorded as 0.94 m. However, the mean calculated for all the jumps, when using a calculator, might give a figure of 0.942857142 m. If you put this number into your results, it implies that you could record the jumps down to a precision in the nanometre range! As a very simple rule of thumb, don't present calculated values to any more decimal places than the original measurements.

What is the best way of presenting the data?

This is often a question students find hard to answer. There is a strong temptation to adopt the 'everything' approach—if you include absolutely everything, then you can't have forgotten anything! However, for the reader this can make interpretation difficult and someone marking your report may conclude that the reason for putting everything in was because you weren't sure what was important and what was not.

Remember that you are telling a story and so try to identify what the key points are, in sequence. Where possible you should present your data in summary form, rather than long lists of raw measurements (which can be included as an appendix at the end if necessary). If you have measured 20 jumps for an individual locust, the reader doesn't want to see a table of the individual jump lengths. Instead, you can simply present the data in summary form as the mean, the standard deviation, and the number of measurements:

> The mean jump distance for the adult locust was 0.87 ± 0.02 m ($n = 20$).

This statement provides the reader with as much effective information as does the full list of individual measurements and is in a much more digestible form. For a practical report, you may wish to include all the raw data, in which case you can add it as an appendix to the end of the report.

The three main ways of presenting data are as follows:

- text;
- tables;
- graphs.

Text Use statements of specific values in the text when you are referring to only one or two items, so the example used earlier, stating the mean jump distance, is appropriate for presenting in text.

Tables Tables are very useful for presenting data in an organized form, particularly where you wish to present several variables together to give an overview of a set of results. The layout of the table (e.g. see 'Table 1') is important to ensure that the contents can be easily read and understood. In particular:

● *try to avoid having too many columns*: having more than five or six columns can make comparison of the data items difficult and also means that the text may be compressed to fit on the page;

● *make sure the rows and columns have clear headings, with units*, so that the nature of the data is immediately obvious;

● *spread the data out* and have clear delineation between rows and columns.

Give your table a number so you can refer to it from the text

The title should allow the reader to understand the table on its own

Each column must have a heading to identify the data set and the units of measurement

Table 1 Measurements of body mass, femur length, and maximum jump distance for five adult locusts

Locust	Body mass (g)	Femur length (mm)	Maximum jump distance (m)
1	1.34	21.4	0.96
2	1.53	21.8	0.87
3	1.65	21.9	0.88
4	1.72	22.0	0.92
5	1.68	22.0	0.97
Mean	1.58	21.8	0.92
SD	0.15	0.25	0.04

Graphs There are many different ways of presenting data as graphs. The types of graph you are most likely to use are:

● bar chart;

● histogram;

● scatter graph;

● line graph.

Bar charts can be used to display data related to frequency. So if instead of displaying the relative proportions of each grouping of locust, you wanted to display the actual numbers in each group, you would use a bar chart as shown in Figure 8.4.

Note that in Figure 8.4 each axis is labelled to identify the information being displayed and that the scale of the *y*-axis (vertical) is set appropriately to display the spread of the data (have a look at the appendix at the end of this chapter for some examples of how *not* to draw graphs). A key feature of the bar chart, unlike the histogram (see next paragraph), is that the bars are separated by a space. This reflects the fact that the *x*-axis (horizontal) is displaying discrete items and not parts of a continuous population. The horizontal axis is just there as a platform for the bars.

Histograms are also commonly used to display data related to frequency, but in this case the variable normally shows a continuous variation. This means that the groups along the *x*-axis

Figure 8.4 A bar chart showing the numbers of locusts divided approximately in terms of age, based on exoskeleton colour (pink: youngest; brown and yellow: oldest), in the sample tested by the class (*n* = 100)

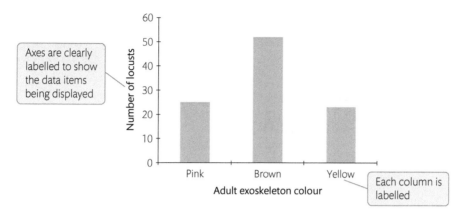

Figure 8.5 A histogram showing the frequency distribution of body masses in the sample of adult locusts tested by the class (*n* = 100). Note that body mass shows a continuous variation and the divisions between the bins are chosen to best display the data.

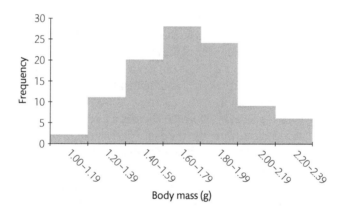

form part of a spread of data (e.g. the height or weight of a population), rather than being separate classes, as is the case for the exoskeleton colour. The example shown in Figure 8.5 illustrates the frequency distribution of body mass for the sample of 100 locusts tested in our class experiment. Here, the body masses of the locusts have been grouped into seven groups (or 'bins'). It is important to select the size of the bins carefully so that you can display the data effectively. Avoid having a large number of bins with only one or two items in each. Likewise, having too few bins may mean that the pattern of distribution of the data cannot be seen. You may need to try a few times before you get the best layout.

Scatter graphs are often used to display the spread of data for two variables that are related to each other. For our sample of locusts, this might be the relationship between body mass and maximum jump distance, as shown in Figure 8.6.

Figure 8.6 A scatter graph showing the relationship between body mass and maximum jump distance for adult and juvenile locusts (*n* = 56)

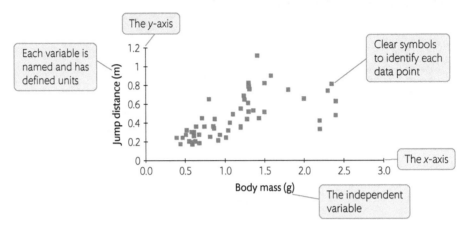

How do you determine which variable goes on the *x*-axis (the *abscissa*) and which on the *y*-axis (the *ordinate*)? The variable placed on the *x*-axis should be the *independent* variable. This means it is the variable that is already known, or can be controlled by the experimenter. In the locust jump experiment we have a group of locusts of known body mass and we want to see how far they can jump. The body mass is the independent variable and the jump distance is the *dependent* variable, which is placed on the *y*-axis. The dependent variable is, therefore, the variable that may change as a function of the independent variable. So, in this experiment, we are proposing that the distance a locust can jump depends on its body mass. The logic of this should be clear: in this case, we can assume that the body mass of a locust does not depend on the distance it can jump!

When displaying data as shown in Figure 8.6, there is a further decision to make, which is whether to draw a line to indicate a trend in the data. For example, in Figure 8.6 it appears that, although the data are spread out, there is a trend of longer jump distances being obtained by the larger individuals. Discussion of the processes of line-fitting is beyond the scope of this book: you should refer to the companion text on statistics (*Research Methods for the Biosciences* by Holmes *et al.*, 2016, Oxford: Oxford University Press). As a rule of thumb, however, be wary of the complex line-fitting programs that can be found in many spreadsheet and statistical software packages. These programs can fit scattered data with very complex mathematical equations, which can be very difficult to interpret. If you think there may be a simple trend such as a linear relationship, then use the program to calculate the linear regression line that best fits the data. However, you should only draw conclusions regarding the linearity of the relationship if you also calculate the coefficient of correlation and this demonstrates a statistical significance.

Line graphs: rather than having a scatter of data points, as for the population in Figure 8.6, your experiment may have involved taking a series of measurements against a set baseline. An example of this would be taking measurements at set time intervals. For such data, time is represented on the *x*-axis and the change in the variable on the *y*-axis. In the case of the data shown in Figure 8.7, the experimenters have measured the power generated by the leg extensor muscles as a function of joint angle. The measures have been repeated for 10 individuals

Figure 8.7 A line graph illustrating the relationship between the mean power output of the locust extensor muscle and the initial joint angle for 10 adult locusts. The error bars show the standard deviations of the means. There was a significant negative correlation between power output and joint angle ($r^2 = 0.91$, $P < 0.01$; method of least squares), as indicated by the linear regression line.

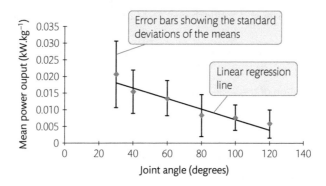

and then averaged, so each data point actually represents the mean of the 10 individual results. This plot therefore illustrates another important feature of graph-drawing which is that, for averaged data, you should also indicate the spread of the data items contributing to each value displayed. This is done by including the standard deviation (see 'Statistics' in the 'Methods' section of Section 8.3.2, 'Get the structure right') as vertical error bars above and below the data point. The reader can then see how reliable the plotted points are in illustrating the trends being shown.

As for the scatter plot, you will need to decide whether to draw a line to illustrate a trend in the data. This can be done either by simply joining the individual points together, or by calculating the regression line that shows the best fit to the data, as has been done in Figure 8.7. Again, you should only plot a best-fit line if there is a statistically significant coefficient of correlation. This coefficient value should be given in the figure legend.

Checklist for graph-drawing

Make sure that:

- you have used the appropriate form of graph for displaying your data;
- the data are displayed on the correct axes—the independent variable is on the x-axis, the dependent variable on the y-axis;
- the axis scales are appropriate for the spread of the data;
- the data points and the axis labels are large enough to be clearly legible;
- the axes are labelled and the units of measurement given;
- the graph has a clear figure legend or title so the reader can appreciate what is being displayed without needing to read the accompanying text.

Look at the appendix at the end of this chapter for some examples of how *not* to draw graphs.

Using images

You might use a variety of images, such as photographs, in your reports. These photographs may be taken using microscopes or electron microscopes, or they may be photographs of a landscape. The key rules here are to make sure that:

- the illustration has a legend or title that explains what is being shown and how the image was obtained;
- any specific aspects of interest are identified by arrows, so the reader can appreciate what is being described;
- there is an indication of the scale of the image—this is usually done by adding a labelled scale bar to the illustration.

Results checklist

After you have completed the writing of your results section, read it through carefully and ask yourself the following questions:

- Have I clearly described the key findings in the text?
- Have I given the correct units for all the measurements?
- Do the tables and graphs present the data in a clear form?
- Are the key conclusions supported by appropriate statistical tests?
- Have I avoided trying to interpret the findings?

Discussion

The discussion is the section of the report where you interpret the findings from your experiment and place them in the context of the research literature. When it comes to writing up a practical class or project, this is often the most demanding part of the exercise, because you must display your understanding of the work you have done and how it fits into the studies undertaken by other scientists. As we mentioned earlier in this chapter, if you have undertaken the practical work in a group, this is where you will be expected to analyse and draw conclusions from your findings *independently* of each other.

A useful starting point for the discussion is to write an introductory paragraph that summarizes the key findings from the results:

> In this investigation into the mechanisms of the locust jump, it has been shown that . . .

This paragraph then provides a good link with the results section and gives you a clear list of the points you need to interpret. There should also be a link back to the aims that you set out in the introduction. Having identified your key findings, you then need to discuss each of them in turn. As for the introduction, you will need to refer to the relevant research papers, highlighting their findings and relating them to what you have found. It may be that your findings are different from, or even contradict, those presented in other papers. In such cases you will need to explain why you think this may be so—don't always fall back on the staple explanation that you probably got something wrong! Whereas that may be true in many cases, there may also be genuine reasons for your experiments having generated different results. If you are writing up a research project, it is likely that you will have new data and these will need careful interpretation with reference to the research literature.

You should remember to round off the discussion with a concluding paragraph that summarizes both the key findings and their interpretation. In class reports, you may also be encouraged to identify what experiments you might do next in order to progress the work. An example of a short conclusion for our locust jump experiment might read as follows:

> In these experiments, it has been demonstrated that there is no significant correlation between the body mass of adult locusts and the maximum distances that they can jump. These findings have been discussed in relation to the studies by Smith and Jones (2014), who reported that the processes of power generation by the adult locust could be more closely related to age rather than body mass. The relationship between . . .
>
> There was a large range of variation in the results obtained in these experiments and any future study should employ more locusts, with more jump trials for each locust, in order to increase the probability of demonstrating significant differences between the different groups. Furthermore, the failure of the heating system during the course of the experiment meant that the later jumps were being performed at temperatures that were sub-optimal for the locusts.

References

As with all your written work, you must include a full list of the references that you cited in the text (see section 4.5, 'Citations and references').

8.3.3 Other sections

You may wish to include some additional sections to the report. The most common of these would be an appendix and a set of acknowledgements.

Although the results section is where you present your data, this is often in the form of processed information, for example, the means of sets of measurements. However, you may wish to include all the raw data, to demonstrate the full set which you have obtained and from which you have derived the summary results. You may also wish to include worked examples of your calculations, such as the statistical tests. Under such circumstances, these can be presented in tables in an *appendix*. This is useful in a practical report, but would not be typical practice in a research paper.

It is unlikely that you will need to include an *acknowledgements* section in a report from a class practical, but this is more common in a project report. For your project you may have been given specific support by academic or technical staff, or been given access to specific items of equipment that would not normally be available to you. It is then a matter of good practice to acknowledge this help in a brief statement at the end of the report. This should not be treated like the eulogies at the Oscar ceremonies—just give brief, factual acknowledgement of any specific support you were given.

8.3.4 Write the sections in an appropriate order

Writing a report can often seem a daunting process, particularly if it is for a large piece of project work. Before you put pen to paper (or fingers to keyboard), check again any brief you have been given for the structure of the report so you know exactly what is expected of you. Also refer back to any feedback you have received on previous reports.

Some sections of the report should be much easier to write than others. For example, the methods section is simply a factual description of what you did and so this is often a good section with which to start.

The next section to tackle is the results. You should think carefully about how you want to present the data first of all, and then produce your graphs and tables, and carry out any statistical analyses. When you have the data presentation sorted out, the process of writing is fairly straightforward since you are aiming to describe the results that you have in front of you. Don't forget that you are telling a story in which the storyline is illustrated by the figures and tables.

Having set out the key findings of the experiments in the results section, you are then in a position to consider their interpretation. Before beginning to write the discussion, however, you should read through the relevant research papers so that you have a clear picture of the relationship of what you have done to the current research. Take each of your key findings in turn and write about it, making sure you explain the relationship with the previous research in order to interpret what you have found.

The final sections to write are the introduction and the abstract. Having written the discussion, you should be in a position to present the background to the work and to set out the aims. As for the discussion, you will need to refer to the relevant research articles to support your statements. This should lead logically into the aims of the experiment and you can check off the aims against your final conclusions. The abstract is the final piece to be written, based on a very brief synopsis of the other sections of the report and highlighting the key findings of the research.

Finally, as with all written work, it is always a good idea to put the completed report to one side for a few days and then to re-read it, checking that it all ties together and could be understood by someone who had not done the experiments.

 Chapter summary

This chapter has highlighted the importance of preparation before a practical class and what to do during a practical class to make the write-up afterwards an easier process. When writing up, it is important to use an appropriate writing style and to adhere to the conventions of scientific report structure. When writing the report it is not necessary to write it in order—in fact there are sections (the abstract being the most obvious example) that can only be written last, and there are other sections that come in the middle (the methods and the results) that are easier to write first. Finally, read through and check your report to make sure that it tells a logical story and that there are no errors or omissions.

Appendix: examples of how *not* to draw graphs

The display of data in graphical form is a key part of the preparation of many reports of practical work, whether it be for enzyme assays or the populations of birds in a cliff nesting site. In the section on graphs we have described some of the key principles in drawing graphs, but there are many pitfalls, especially when using spreadsheets or graph-drawing packages. Some of these pitfalls are more obvious than others. Have a look at the graphs that follow (Figures 8.8 and 8.9) and list the errors that you see: each shows at least one error!

Figure 8.8 shows a number of common errors (although rather exaggerated) that might be made when displaying the data shown in Figure 8.7:

Figure 8.8 Power output against joint angle, locust experiments

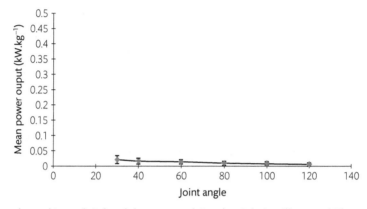

- *The figure legend is too brief and does not explain what is being illustrated.* The test here is: if the reader only had the figure and the legend to go on, and no accompanying text, would he or she be able to understand what was being shown?

- *The scale for the y-axis is inappropriate.* The vertical spread of the data points does not even extend to the first scale point on the axis. Furthermore, the compression means that the error bars are not actually visible. The scales should be set so that the data points extend over almost all of the length of the axes.

- *The font size for the scale on the x-axis is too small to be legible.* Make sure that the fonts for the two axes are legible and are the same size.

Figure 8.9 Scatter plot showing the relationship between body mass and maximum jump distance for adult and juvenile locusts, *Schistocerca gregaria* (*n* = 56)

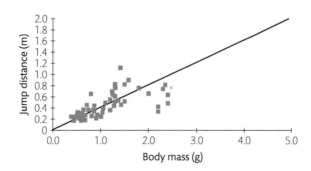

- The x-axis label 'Joint angle' does not specify the units.
- *The numbers on the y-axis are not set to the same level of precision.* Make sure that the same number of decimal places is specified for the scale indicators.

Figure 8.9, at first sight, might appear to be okay, since the axes are appropriate and clearly labelled, and the legend informs the reader as to what data are being presented. However, there are some key flaws in the presentation:

- A straight line has been drawn through the data points, but there is no indication as to the basis for drawing such a line: was it simply drawn by eye as appearing to be the line that best fitted the data, or was it calculated using linear regression? The method for arriving at drawing such a line needs to be stated in the figure legend. If it is drawn by linear regression, then the method chosen for calculating the line needs to be stated and the coefficient of correlation given so that the reader knows whether the linearity is statistically valid. An example of the statement to include in the figure legend would be as follows:

- Is it valid to extrapolate or interpolate a line? The straight line here has been drawn back to the origin (coordinate 0,0); this is *interpolation* and it makes the assumption that the linearity of the relationship continues below the range of the data set on which the calculation was based. Therefore, you need to ask yourself whether this assumption is justified or in this case biologically possible—after all, how many locusts are there that weigh 0.01 g? The same argument holds true for *extrapolation*, the process of extending the theoretical relationship beyond the range of the data set. Do you know of any locusts that weigh in at 5 g? Apart from the fact that they would be big and scary, can you be sure that they could jump 2 m? The rule of thumb is that if you draw a straight line through the data set, it should only extend over the range of the data points and not beyond, unless you have strong evidence that the relationship will hold true for other data points.

9 Working in tutorials and groups

 Introduction

Tutorials and group work form key elements of teaching within most degree programmes and can be employed in a variety of formats, as we will see in this chapter. Irrespective of the specific format of a session, the most important things you can do are to *prepare* and to *participate*. Without preparation, and if you are not willing to participate, you will derive little benefit from a tutorial or group work session.

Tutorials can be quite a daunting prospect, as many students are unsure of their knowledge base and ability to discuss ideas, particularly in front of a member of staff who may be a world expert on the subject. But you should remember that your fellow students will be in a similar position and your tutor will want to encourage you to participate.

Tutorials and group work are useful in developing a range of important skills that will be of value to you in your studies and also in your career development. These skills include the development of:

- *subject awareness*—increasing your knowledge of the subject;
- *critical thinking*—listening to different ideas as they are put forward, thinking about them, and evaluating them in terms of the strengths and weaknesses of the arguments;
- *communication skills*—presenting your own ideas in a way that can be understood by the rest of the group, and learning the essential ability to be able to refine the views expressed by other people, pick out the key points, or even counter them, without causing offence.

These skills, particularly the critical thinking and the communication skills, link closely with the skills that are sought by graduate employers (Chapter 15, 'Making yourself employable'), so developing them is beneficial both for your immediate academic performance and also for your future career prospects.

We will begin by addressing tutorials.

9.1 Tutorials

The basic format for a tutorial is a relatively informal group, where a tutor meets with a small number of students to discuss specific aspects of the course or module being studied.

9.1.1 **Different types of tutorial**

There are two main types of tutorial.

- Academic tutorials that are normally linked to the modules you may be taking at the time—these sorts of tutorials may, very occasionally, be run on a one-to-one basis, but will more commonly involve five or six students, or even up to 20 or so. Academic tutorials may involve a range of activities such as discussions about a specific topic, short presentations given by members of the group followed by a discussion, or the consideration of the answers to previously circulated questions.

- Tutorials with a personal tutor, or equivalent, to discuss your progress on the course— these will often be on a one-to-one basis and may include discussion of your progress in your studies, your career plans, and personal development planning, as well as aspects of the extra-curricular areas of university life, such as your engagement with societies.

As well as tutorials, you are very likely to be required to participate in various forms of group work. Many of the principles, particularly in terms of engagement, are the same for the two types of session, but we will also consider group work separately in Section 9.2, 'Group work').

9.1.2 **Preparing for tutorials**

Preparation for tutorials is essential. It may well be that you have been given a specific topic to research, a worksheet to go through, or some sample questions to answer. On the other hand, you may simply have been told what the theme of the tutorial will be. Either way, you need to be prepared, so that you can benefit from the discussions and participate in a meaningful way.

Preparing for academic tutorials

Academic tutorials offer you the opportunity of discussing your subject with a bioscientist, who is very knowledgeable, as well as with your fellow students, in a fairly informal setting. If you are going to be able to engage in that discussion and to learn from it, then careful preparation is essential. It is likely that you will be given one or more specific topics that will form the starting point for the tutorial discussion. As with the preparation for writing an essay (see Chapter 7, 'Writing essays'), or giving an oral presentation (see Chapter 10, 'Preparing scientific presentations'), you need to read up on the topic and make succinct notes that you can use as prompts in the discussion.

If the tutorial topic is based on material that you have been covering in your lectures, then the obvious place for starting your reading will be your lecture notes, before moving on to read any specific references you may have been given. Have a look at Sections 4.1.1, 'Textbooks', 4.1.5, 'Research papers', and 4.3, 'Note-making strategies'. As a guide, the most important things to ensure are that:

- you are keeping focused on the question asked;
- you are making brief notes that summarize the key points and which you will be able to use as prompts in the discussion;

- you are keeping a record of the sources of the information (see Section 4.5, 'Citations and references') so that you can cite the paper or text if you are challenged regarding the reliability of the information you are presenting, or if you want to find the information again to get more detail.

A short time before the tutorial, read through your notes to make sure that you understand the points and to refresh your memory.

Preparing for personal tutorials

Personal tutorials may cover a range of topics. In relation to the themes of this book, the most important aspects are likely to be topics such as your academic progress, module choices, personal development planning, and thoughts on careers. Again, if you are going to benefit from the opportunity, then you must prepare in advance and be ready to discuss these issues openly.

For example, in order to be able to discuss your current progress, it would be very useful to take the marks and feedback you have received from recent pieces of coursework with you. Re-read these and check, for example, if there are similar comments that have been made on several different pieces of work, which would be a good focus for attention.

To discuss your proposed module choices for a forthcoming semester or academic year you need to have read about the range of choices that are open to you. This will include checking whether taking the modules requires you to have taken specific prerequisite modules in the current year of study. You also need to have thought carefully about where your strengths and interests lie, and indeed, if you have any particularly weak areas of the subject that are best avoided. In the context of your development, you should also be thinking about whether you would benefit from further training to develop your skills. For example, if the modules are assessed by poster presentations and you are not sure how to go about preparing posters, then you should read Chapter 12, 'Creating academic posters', and also consider seeking guidance from your tutors or perhaps from the university's learning support centre, or equivalent. Finally, you should also be thinking carefully about where you think studying those modules might take you in career terms, with regards not only to the subject content, but also the specific skills that you will be developing (see Chapter 15, 'Making yourself employable').

Try this—Preparing for a personal tutorial

This is a useful exercise, even if you are not scheduled to be meeting with your tutor in the near future, as it gives you the opportunity to take a step back and reflect on where you are in your programme. Try writing a sentence or two in answer to each question:

- Am I enjoying the course or are there some aspects that I enjoy more than others?

- How am I coping with the different modules and how am I dividing my time between them?

- Are my marks as good as I expect/would like?

(Continued)

- Have I read through the feedback I have received on my work and have I acted on it?

- Are there any specific aspects that I am struggling with (this could be specific topics or activities such as essay-writing)? If so, do I know where to get help?

- What are my plans for the next set of modules or for when I graduate? Am I taking the right modules at the moment to help me transfer to the next stage?

- Are there any personal issues I want to discuss with my tutor or with other university support services?

Do try this just before a tutorial, but also around the mid-point of the term; it is good practice to reflect on how things are going.

9.1.3 During tutorials

You have come to the tutorial, having done your preparation. The next important element is to make sure that you participate. Participation, of course, means making active contributions to the discussions, but it also means listening carefully, thinking about what is being said, and relating that to what you have already learnt.

Contributing to academic tutorials

The first and most important thing to remember about tutorials is that they are not mini-lectures: your tutor will expect to have a role in facilitating and guiding the discussion, but will not expect to be talking for much of the time. Indeed, some tutors are experts at sitting in silence waiting for the students in the group to start talking! There will, therefore, be an expectation that you will be an active participant—that you will attend the tutorial having prepared beforehand and be prepared to discuss your ideas, as well as to listen to the points made by your peers and the tutor.

You will probably find that this is all quite a daunting prospect at first and there is a strong temptation to sit looking at the floor, hoping not to be noticed! While this is an understandable response, it is not a very productive approach. Remember that your tutors are there to help you understand the subject and will want to help you develop your skills in communicating ideas; however, they can only do that if you participate. Don't be afraid of expressing ignorance if there is something that you don't understand—the only thing that is likely to be irritating to the staff is if it is apparent that you have not bothered to do any reading or preparation beforehand.

Depending on the structure of the tutorial, there may be some short presentations about topics that you and your peers have been asked to prepare in advance, followed by discussions. Alternatively, there may be specific questions that you have been asked to discuss, so the whole tutorial takes the form of a facilitated discussion. Irrespective of the format, there are key ways of engaging with the process.

- Listen carefully to any ideas being presented by the tutor or other students, and note down the key points of the arguments. At the end of the statement, say whether you agree or disagree with the ideas, presenting evidence to support your views. It is no good just saying you agree or disagree with an idea—you also have to be able to justify that view.

- If there are things you don't understand, then ask for further details or explanation. Never leave the tutorial not understanding something because you were afraid to ask questions.

- At the end of the discussion, make sure you can summarize the key points for yourself.

- You do need to be sensitive, though: if you are a confident speaker, don't try to dominate the discussion—let other people express their views as well, since that is an important way of learning.

Also, though, if someone else is dominating the discussion and you don't feel you have the chance to speak, then make sure you attract the tutor's attention, for example, by raising your hand.

Contributing to personal tutorials

With personal tutorials, as with academic tutorials, you can only expect to benefit if you engage actively in the process and, in these cases, are prepared to be open in terms of your views and aspirations, as well as being realistic in terms of your abilities. This means that you may need to open up discussion topics and explain your background to your tutor, rather than waiting for the topics to be raised. Despite the best of intentions, your tutor may not be able to guess what is on your mind or worrying you at the moment. You should also recognize that your tutor may not have all the answers, although, they should be in a position to be able to advise you on where to seek further guidance.

It is important to remember that your personal tutor is there to try to help you. Often students are anxious that if they raise issues of concern with their personal tutor, it will affect how the department views them. This should not be the case under any circumstance. Although the personal tutor is not bound by a confidentiality clause as are, for example, GPs or counsellors, they have a duty to be discreet about things they are told and not to be judgemental.

If you have issues that concern you or you want to seek advice, don't be afraid to make an appointment to see your tutor. Some departments operate an 'office hours' policy, whereby tutors will always be available at specific times during the week. Other departments may have an 'open door' policy where you can go to see a member of staff at any time. Either way, it is often a good idea to contact your tutor first to make sure you can see them at a specific time that fits in with your timetable. Scheduling a meeting also means that your tutor should have time for a proper conversation.

9.2 Group work

The process of collaborative or group working may often be very productive. Indeed, there is a wealth of evidence that material learned through group work is retained better than that learned through formal teaching methods. Group work is often very useful in developing critical skills because, as a member of a group, you need to evaluate the contributions that your fellow students have made, as well as reflecting on your own contributions and discussing their validity with the rest of the group. In the biosciences, group work is very common as a part of laboratory classes, where you typically may have to work in pairs or threes in order to complete the experiment. In the case of group work based around discussion questions, many of the points relating to tutorials may also be applied. In particular, you again need to be thorough in your preparation and be prepared to engage actively in discussions.

It is very important to be clear about the extent of the collaboration, for example in a practical class you may be required to work together to obtain the data, but then to write individual reports based on that data. In another exercise, you may be asked to work together to produce a written report, but also draft out individual reflective commentaries on the process. In these types of exercise, be very clear where the boundaries lie between the collaborative and the individual elements, so that you don't fall foul of your university's regulations on collusion (Section 5.1.3 Deliberate plagiarism, Collaboration and collusion). For the purposes of this chapter, we will assume that you are being given fully collaborative exercises to undertake and that there will be a single product, produced by your group, as the outcome.

The organization and dynamics of interaction of your group will be critical to its success, and will be strongly dependent on everyone being willing and able to support the group as a whole. Unlike the tutorial, where your discussions are facilitated by a tutor, group work exercises will normally require you to work with your group on a set task, perhaps in a classroom setting for a relatively short time, or as an extended exercise that you undertake at times and locations of the group's choosing. In some exercises, you may be required to work online without any physical meetings: this may be through an online conferencing tool such as Skype or Zoom, where you can see each other, or it may be through text-based interactions where you have to follow threads. However you are engaging with the group, the dynamics of how people interact and contribute to the group are a very important element contributing to the overall success of the group. Under these conditions, how you behave in the group is also clearly important. Some key points to remember here are:

- *keep an open mind* and be prepared to listen to, and consider, the ideas put forward by others in the group;
- *avoid being judgemental*—for example, by being dismissive of someone else's ideas;
- *be careful of your body language*—for example, folding your arms and staring out of the window does not convey a positive message;
- *if you are engaging virtually through text alone, take care how you phrase your statements*—when you are trying to express views succinctly through text, it is often also very easy to come across as being very brusque.

Let's now consider each of these types of group activity.

9.2.1 Short-term group work

Short classroom exercises are often employed as a way of breaking up lectures or as a means of checking that students have understood a specific concept. You will probably be asked to talk to the people around you to discuss a specific topic or answer some questions. Many of the issues of group dynamics and team work that are discussed in section 9.2.2 'Longer-term group work' are not applicable here. For the most part, the most important skill is remaining focused on the topic and using the time constructively—it is very easy for discussions to degenerate into a review of the latest gossip and, although that might be of great interest, it will not help get the task completed! It is also often very easy to spend a long time debating the first one or two aspects of the question and then find you have run out of time before completing the exercise, particularly if you only have a few minutes for your discussion.

There are several ways of providing some structure to your discussions and making sure that you get some answers, while also getting everyone involved as far as possible.

- If you have been given several questions to answer, it is very helpful to agree quickly at the start how long you are going to spend on each question, and make sure that you adhere to that timescale, noting down your agreed answers as you go.

- A useful way of getting ideas quickly and of engaging everyone in the discussion is to go round the group in turn with everyone putting in an idea or point of information related to the question.

- An alternative is to generate ideas. People in the group mention ideas in any order as they think of them. This can be very productive if some creative thinking is needed, because the lack of formality encourages more open thinking. It can also generate numerous ideas very quickly. However, it may also lead to some people feeling excluded because, for example, they feel apprehensive about the merit of their ideas or because they are too shy to contribute. In such situations, try to be aware of the other members of the group and encourage non-contributors to participate: after all, they might have the best ideas of everyone! As with the guidance for behaving in tutorials, the key points here are to listen to what others have to say, to ensure your body language is appropriate and open, and not to be dismissive, even if you disagree with what the other person is saying.

9.2.2 Longer-term group work

If you have been allocated a task that will require you to work in a group over a period of days or weeks in order to complete the task, then the dynamics of the group become a very important factor, contributing to the success, or otherwise, of the exercise. Many students are dubious about the benefits of group working, particularly if the exercise is worth a significant amount of marks. However, the skills involved in such exercises are very important for career development, because team working skills are among the most highly valued by employers. When you get to the stage of applying for jobs and being interviewed you will doubtless be asked questions about your ability to work in a team, what kind of role you play, and to give examples of when you've had to work hard in a team context to complete a shared task. Pointing to experiences of sports teams, or your involvement in clubs and societies, or part-time work are all good examples, but being able to give examples from an academic context is important, too.

Undertaking a specific project as a group can be broken down into a series of key stages, as shown in Figure 9.1. We will consider each stage in turn.

Get off to a good start

It may be that you have been allowed to select who you work with and have chosen to work with a group of friends. On the other hand, you may have been put into groups by your

Figure 9.1 Stages in the process of group work for a specific project

lecturer, and you may or may not know the people you are having to work with (and, if you do know them, you may not like them!) You might not feel happy if you find yourself grouped with people you don't know, or perhaps don't like, but it can be very good experience for your working life, since you may well have to team up with strangers to undertake a task in your employment, or you may have to work with people you don't necessarily like. In either case, the work still has to be done.

Time spent in ensuring that the group gets off to a good start should pay dividends as the project develops. Setting up formal structures for the group—like having a group leader, agreeing task allocations, or having meeting agendas—might seem rather over-the-top for some projects, but it can all be useful experience. Additionally, having formal aspects to the project, which are agreed in advance, helps to avoid problems later on.

Get to know each other

This first stage is clearly most important if you are going to be working with people you don't know. Even in a group where you know everyone, that might not be the case for everyone else. So, it is very helpful to get together for your first meeting, away from the work environment, perhaps in a coffee shop or pub, or just sitting on the grass if the weather is good! Use that meeting to ensure that everyone feels part of the group and can, at least, begin to feel confident about working together. If you need to do a round of introductions, get each person to say something about themselves as a 30-second biography, or say what they think might be particularly interesting about the project.

Establish your ground rules

Don't be misled into thinking that, just because you are working with friends, you can all just get on with the task. Whether you are with friends or strangers, it is essential that you agree the ground rules for running the project, otherwise things can come unstuck very quickly. The extent of these rules will depend on how long the project will be running for. If it is relatively long, for example several weeks, you might like to consider the following questions/points, which could be set out in a generic agenda for the first meeting (as shown in Figure 9.2):

- Do we need a project leader/coordinator who will keep track of progress?
- What technology shall we use to store our joint work? Local files on individual computers are not collaborative and may make the group over-reliant on one individual. Consider using more collaborative software like Microsoft SharePoint or Google Drive, where 'live' files can be accessed and edited by the group members (other examples include Dropbox or your university's virtual learning environment).
- Would someone take the role of secretary to keep a record of points agreed at meetings (as shown in Figure 9.3), and make sure that meetings stick to time? (It doesn't have to be the same person each time.)
- Do we need formal agendas for future meetings? If you think that setting agendas makes the whole process seem too formal, then you could have meeting plans.
- What time commitment should each member of the team be making to the project?
- What action should be taken by the group if one or more members are not contributing?

Figure 9.2 Proposed plan for the group's first meeting

BioStars Project Group Meeting Plan
14:30 Tuesday 2 May
Biology dept coffee bar

1. Introductions
2. Group organization
    a) Do we need a group leader?
    b) Do we need a secretary?
    c) Do we need . . .
3. Group operation
4. Brainstorm
5. Division of project tasks
6. Key progress dates
7. Next meeting

Figure 9.3 Example of first section of notes from the group's first meeting

BioStars Project Group 1st Meeting Report
14:30 Tuesday 2 May
Biology dept coffee bar

1. Present
Alex, Komal, Sarah, Lee, Rathan

2. Organization
Komal will act as group leader, Sarah will take notes

3. Operation
Project task is to present a critical review of deep brain stimulation in the treatment of Parkinson's disease. 6 weeks working time.
Group to meet every week to update

4. Brainstorm
PD – degeneration of dopaminergic neurons in substantia nigra (basal ganglia) movement deficits . . .

Generate ideas

You have been given the project title and a brief for what is expected of you. No doubt different members of the group have different ideas about what the approach should be, what needs to be done, and how. Use a free, unstructured session to let everyone have their say. It is important that the discussions at this stage are free-ranging, so that the full breadth of the topic can be covered and also so that everyone feels free to participate. At this stage, all contributions should be treated equally and there should not be any criticism of the ideas that are offered: you will have time later on to focus on specific aspects.

(*Note*: it is important that someone records the ideas, otherwise all that brain power will be going to waste! However, you need to make sure that the person recording the ideas has the chance to contribute ideas as well and is not just there as a note-taker. Also, set a time limit for the discussion, otherwise you could be there a long time.)

A useful approach to generating ideas is to give everyone two or three minutes to think about the problem and then go round the room so each person puts their ideas in. These ideas can be recorded as bullet points on a flip chart, which everyone can see. When the first round of ideas is up, see if there are links between them so that you can group them together. From that grouping, you can then go round the room again to build on these with further ideas, but *remember your time limit*!

Agree tasks

Once you have the outline of how to approach the topic, you need to break this down into a series of tasks, so that you can plan the way the work will be done and who will do it (as shown in Figure 9.4).

Figure 9.4 Identification of the tasks involved in preparing the presentation

Tasks for Parkinson's disease project

Research topics

1. Structure & function of the basal ganglia
2. Neurological features of Parkinson's disease (PD)
3. Current drug treatments
4. What is deep brain stimulation? (DBS)
5. Use of DBS in treating PD

Identify the key points for the presentation
Prepare the presentation and write the text
Prepare handouts for the audience
Rehearse and review the presentation

One of the obvious features of group projects is the division of labour so that there is co-operation; if you have everyone trying to do everything, there is no real benefit from being in a group. On the other hand, you need to make sure that everything is covered. So it is vital that, right from the outset, the group agrees who is going to do what and that a common record is kept of the division of labour (Figure 9.5). So, having drawn up a list of tasks from your ideas session, the group needs to agree how they are divided up so as to be reasonably equitable. At this stage, it is really helpful to take account of different people's particular skills—if you have a successor to David Attenborough as a presenter, then it makes sense for that person to be taking a major part in any presentations.

Figure 9.5 Key dates for the project and breakdown of project tasks

Project timeline:

May 9th	Completed background research
May 16th	Agree the bullet points for the presentation
June 6th	Draft presentation complete
June 10th	Rehearse presentation
June 13th	Class presentation!!!

Work division:

May 9th

      Komal – neurological aspects of PD
      Alex – current drug treatments
      Rathan – what is deep brain stimulation?
        . . .

Plan your timeline

As well as dividing the project up into different elements that individuals will be responsible for, you also need to divide it up into chunks of time so that you have specific milestones that you can identify, in order to record progress. It may be helpful to link this in with the agreed division of labour, so that each person has a target to work towards and take responsibility for (Figure 9.5). Also, build in some slack in the timing, so that you allow some extra time at the end of the project for completion. For example, you could aim to complete the project a week before the deadline, then if there are any problems you do have some leeway.

> Try this—Practise planning
>
> Think of a simple task with several steps that you are going to carry out by yourself, such as preparing a meal for yourself and three friends. Decide what time you want to eat and write down each step required with the time it needs to be done by. Repeat the process, but now assume that your friends are helping you to prepare the meal. Write down each step involved, the time it has to be completed by, and by whom.

Keeping going

It is often the case that everyone will be pretty enthusiastic about the project to begin with; the first meeting goes well with agreed plans of action and so on. Keeping the momentum going, however, can be much harder. This is especially true when more imminent coursework deadlines focus input elsewhere and the long-term nature of the group project tends to give rise to an element of complacency. This is when it becomes important to monitor—and stick to—your original project plan.

Meet regularly

Keep to your schedule of regular meetings. In particular, don't let a meeting be dropped because people have not done their work (even if one of those people is you!); it is very easy to let this happen, and you will quickly find you are on a slippery slope and no progress is being made. If nothing else, therefore, the meetings will act as a reminder that the work needs to be done and that the group needs to keep on schedule, otherwise the project will be compromised. This also adds incentive because people will be embarrassed to admit to the group that they have not done their work.

Keep a check on progress

At each meeting, it is important to review exactly where the project is and whether you are on track. Refer back to your timeline (Figure 9.5) and check off your progress against the target dates.

This checking, however, is not just a tick-box exercise, accepting that the work has been done without being sure that it actually has. If the group is aiming for high quality in its performance, then there also needs to be an evaluation of the quality of the work each person has produced. If your tutors gave you a marking scheme or equivalent, then the members of the group need to evaluate each section of the work that has been produced against the marking scheme. Although it may seem like it, this is not just a chore—it is important for your own learning that you have an understanding of the whole project, not just the sections you are responsible for. It is also a very valuable learning exercise in developing the skills of critical evaluation, reflection, and discussion.

People who fall behind

One of the biggest, and most common, problems with group work is dealing with group members who don't keep up with the work, or fail to engage with the group. There are various strategies the group can adopt to try to resolve the situation.

- *Talk to the person*: this is the first and most vital approach. If the person comes to the meeting, then you can discuss the situation. If he or she has stopped attending the meetings, then you need to try to re-establish contact as soon as possible.

- *Find out what the problem is*: it may be that the person has been ill, or is struggling with something else and can catch up, particularly if given a bit of time or some help as a one-off. If there is no real excuse, then the group needs to refer back to the ground rules and make it clear that everyone needs to provide their agreed contributions, otherwise the whole project may fail.

- *Persistent non-cooperation*: if one member of the group will not cooperate, even after you have discussed the position, or if that person refuses to respond, then you need to take more explicit action. Probably, the best option here is for you to contact the tutor and explain the situation, and ask for assistance in resolving the problem, although this should be an option of last resort as it indicates that the group can no longer sort out its own problems.

The final product

The deadline is approaching and you now need to bring everything together ready for submission. If you have let things slip, then the quality of the final product may be lacking. If, however, you have kept to your timeline, this phase of the project should allow you to produce a really polished piece of work.

- *Bring everything together*: one of the difficulties of group work is that everyone has their own style of writing or expressing themselves, but the final piece of work needs to appear as a consistent piece, not as three or four completely separate elements that have been lumped together. This may require you to edit the sections so that the approach or style of writing is the same and also so that there are sensible links between the separate sections. Editing by committee is often a lengthy and unsatisfactory process. In most cases, it is preferable for one person to be responsible for bringing the sections together and editing them to create a single 'voice'—think back to the initial process of agreeing the tasks so that you could play to the strengths of each member of the team.

- *Rehearse your presentation* (see Section 10.7, 'Practice'): if your final product is a presentation, then it is essential that you rehearse. In particular make sure that:

  - the presentation keeps to time;

- the illustrations all show up appropriately—this is especially important if you are using animations or, for example, showing video clips that are embedded in the talk;

- if you have more than one speaker, make sure that the links between them are smooth and everyone knows who is speaking in turn.

● *Check against the brief*: before you submit the work, run through the instructions you were given at the outset again, and make sure that you have addressed all the criteria. It is important that, as far as possible, everyone agrees that they are happy with the final product and are prepared to sign it off.

When it's all over don't forget to make use of the feedback (see Chapter 2, 'Using feedback'), and think about where any problems were and how you might do it better next time.

 Chapter summary

This chapter has looked at ways of working for tutorials and group work. The key message for all these activities is that you need to prepare, to participate, and to listen to others.

Preparing for academic tutorials means reading through any lecture notes or reference material to make sure that you know the subject well beforehand. For both academic and personal tutorials, you also need to think about what the objective of the tutorial is. During the tutorials, make sure you listen to what others have to say and be prepared to take their ideas on board, as well as argue against them if necessary. Also, remember to take notes, so that you have a record of the discussions.

For group work, you must be prepared to work with other people, even if you don't like them! Agree the tasks to be done, who will do them (work to each person's strengths), and by when, so that the group can keep a track of how the project is developing. To do this, you must keep meeting regularly and checking progress against your timeline: this is particularly important if the project is running over quite a long period.

10 Preparing scientific presentations

 Introduction

The ability to present information verbally to a group of people is an increasingly important skill for students undertaking bioscience degrees, or any other kind of degree for that matter. It's also one of the key skills that many employers look for across a wide range of roles. Some students feel the prospect of standing up in front of a group of their fellow students and academic staff to give a presentation is the stuff of nightmares, even more so when the presentation is assessed. For other students, presentations are not quite such a frightening prospect, but even so, the public nature of a presentation, compared with the privacy and anonymity of an essay or exam, for example, is a factor that makes getting presentations right all the more important; no one likes being embarrassed in front of their peers.

This chapter begins by considering the purpose of presentations as a form of assessment; we will then identify some of the characteristics of effective presentations, scientific presentations in particular. We will look at the importance of finding out the answers to a number of key questions before progressing on to analysing the brief, researching the topic, and planning and ordering your presentation's content. Finally, we will highlight the importance of preparing effective and appropriate notes. This chapter complements Chapter 11, 'Delivering scientific presentations', which looks specifically at delivering, as opposed to preparing for, your scientific presentation.

10.1 What are presentations supposed to achieve?

It can be useful to think about why you are being asked to undertake a particular form of assessment, as different forms of assessment are chosen for different reasons. The assessment of presentations is usually divided into two broad areas: content, and the communication of the content (or 'substance' and 'style'). Both of these elements are important and you need to be careful not to focus on one at the expense of the other. For example, it is possible for your presentation to have relevant and informative content, but if you don't communicate it in a well-structured and engaging manner you would score relatively low marks in an assessment. Equally, you should avoid trying to dress up content that you don't understand and haven't put much effort into researching by giving a slick presentation with a few impressive slides. If you do this, it is usually fairly obvious to the audience (and it certainly will be to your tutor) that your presentation is all style and no substance. So, both substance and style are important.

We will consider the following brief for a sample presentation:

> Prepare and deliver a five-minute oral presentation on the use of global positioning systems (GPS) in the study of migrating populations. The presentations will take place at 12 noon on Friday 1 March in room 101.

We can identify a list of the skills being exercised in order to complete the task as follows:

- using feedback from previous presentations to produce a better piece of work;
- analysing the question or title (and associated instructions);
- researching the topic;
- bringing together the information gathered from a range of sources, e.g. lecture notes, textbooks, research papers, web articles, etc.;
- planning the information to address the brief and makeing sure that all the key points are covered;
- matching the level of detail against the type of audience you will be addressing;
- ordering the material in a logical manner;
- creating visual aids to support your key points;
- preparing appropriate notes so that you can remember what you want to say;
- practising your presentation beforehand so that you become familiar with communicating the content;
- delivering your presentation in a relaxed and engaging manner;
- referencing the information correctly to acknowledge the original sources;
- planning and managing your time to make sure everything is ready by the deadline.

As was similarly identified in Chapter 7, 'Writing essays', this is a surprisingly long list and could probably be extended. The important point is that, while some of these skills, such as researching the topic and preparing your notes, are obvious and easily recognized as part of the process, others, such as using feedback and referencing information, are more implicit, but still very important if you are to plan and deliver a good presentation. As we go through the process of preparing a presentation (this chapter) and delivering a presentation (Chapter 11, 'Delivering scientific presentations'), we will consider these different skills and give you guidelines for improving the quality of your scientific presentations.

10.2 What makes a good presentation?

What, then, makes a good presentation? Nearly everyone has an opinion on this question. Even if you have never given a presentation before, you will have seen others give presentations (for instance, a lecture), and you will probably remember some that were really good, and some that were really bad! It is useful to think about what makes a good presentation, as it will give you a list of things you want to try to copy and things you want to try to avoid.

Try this—Identifying good and bad features of presentations

Identify two presentations that you observed fairly recently, one really good one and one poor one. Consider each in turn and make a list of all the main features of each. Use this list to produce a brief guide of 'do's and don't's' for presenting.

10.2.1 Characteristics of effective presentations

Obviously, not all presentations should be the same. In fact, it is important that, as you gain experience in presenting, you develop your own style, rather than trying to conform too rigidly to a set standard. However, effective presentations usually exhibit certain characteristics. These include that the presentation is:

- relevant;
- informative;
- well structured;
- engaging.

This list is not exhaustive, though these are important characteristics of effective presentations. You should also recognize that these characteristics are not self-contained; they overlap and are dependent on each other. To go back to our substance and style distinction that we mentioned earlier, relevance and informativeness are largely to do with substance, whereas structure and engagement are largely to do with style. We will deal with each of these characteristics in turn.

Relevant

The starting point for any presentation is the presentation brief, as it is this that will govern the content and focus of the presentation. Although it can be tempting to jump straight from a brief to the bits you find interesting about the topic or the bits that you happen to have found some useful information about, it is important that you keep going back to the title and instructions you have been given to make sure that your material is relevant, just like you should do for an essay. However, a presentation should not only be relevant with regard to the brief; it should also be relevant with regard to the audience. Are you pitching your presentation at the correct level for the audience in terms of their current understanding and their ability to assimilate the material? For instance, is it likely that you will be using quite technical terms on occasions? If so, is your audience already familiar with these terms, or do you need to define them? So, a presentation needs to be relevant, both to the brief and to the audience.

Informative

Not only should a presentation be relevant, it should also be informative. What is the content that you want to communicate to your audience? What do you actually want to say? It can be very helpful during your preparation to try to summarize the content of your presentation in a single sentence, or three or four short bullet points. For instance, to go back to our previous example, you could summarize the content of that presentation as follows: 'Short explanation of what GPS is, how the technology has been applied to study migrating populations, and examples of wildebeest and Canada geese.' If you can't summarize the content in a sentence or two then it probably isn't focused enough to be informative. So, a presentation needs to be informative.

Well structured

Now we start to get on to the style, as opposed to substance, elements. The structure of a presentation is a good example of where there is significant overlap between style and substance: good structure begins with the content. However, you need to be able to communicate that structure to the audience so they can follow what you are saying. If a presentation is easy to follow, it will in part be due to its good structure. There are two simple ways of communicating the structure of your presentation to your audience—you could tell them or you could show them (or both).

Telling the audience the structure involves outlining it verbally at the beginning, for instance:

> This presentation is divided into three main sections. First, I'll give a brief explanation of what GPS is. Then I'll explain how the technology has been applied to the study of migrating populations. Finally, I'll tell you about two specific examples: wildebeest and Canada geese.

In addition to outlining the structure of the presentation at the beginning, it will also be helpful to give verbal cues to the audience as you go along so they can clearly follow the progression of the presentation. Verbal cues act like signposts to the audience so they know where you are going. Some of these verbal cues include:

- a single word such as 'First', 'Secondly', and so on;
- linking statements such as 'So that's the background to the technology, let's now think about its application . . .';
- concluding statements such as 'In conclusion . . .' and 'Finally . . .'.

That might all seem fairly obvious to you, but it is surprising both how often these signposts are neglected and how helpful it is for the listener when presenters make a point of using them. As you become more experienced in presenting you will give these verbal cues much more naturally, but for as long as they don't come naturally it is a good idea to make a conscious point of using them.

Showing the audience the structure could simply be a case of producing an outline slide that summarizes the main points. An example is shown in Figure 10.1. You could use this at the beginning when you tell the audience about the structure. If the structure was particularly important or particularly complex, you might also repeat the outline slide within the presentation when you mention the verbal cues. However, it is possible to give a presentation so much structure that it becomes very separated out and doesn't flow, so be careful not to overdo it.

Figure 10.1 Outline slide showing presentation structure

Outline
- What is GPS?
- Applications to migrating populations
- Examples
  - Wildebeest
  - Canada geese
- Conclusion

Engaging

The final characteristic of an effective presentation that we have identified is that it should be engaging; it should be interesting and hold the audience's attention. Again, there is overlap here with the content or substance aspects of the presentation (particularly the relevance to the audience), but whether or not a presentation is engaging is much more to do with how information is communicated as opposed to what is communicated. Whether or not a presentation is engaging will depend on a wide range of factors: it will need to be well paced, the presenter will need to have good eye contact with the audience, and it will probably help if the presenter can be fairly relaxed (or at least seem to be relaxed!). All of these depend to a large extent on how much you have practised the presentation. Additionally, an engaging presentation is often also helped by the appropriate use of visual aids, which we will address in Chapter 11, 'Delivering scientific presentations'.

10.2.2 Characteristics of effective scientific presentations

In addition to the characteristics of effective presentations highlighted previously, there are certain characteristics that scientific presentations in particular should exhibit. These include that a scientific presentation should be:

- objective;
- precise;
- clear.

Again, this is not an exhaustive list and there is overlap between the different elements, but they are key characteristics nonetheless. We will address each of them, briefly, in turn.

Objective

Science is, by its very nature, objective. That is to say, observations made during scientific research should be impartial, and not influenced by personal feelings or opinions. So it follows then that a scientific presentation should also be objective. This means that the content you are communicating needs to be well researched and well referenced. As was highlighted in Chapter 5, 'Academic integrity and avoiding plagiarism', students are usually aware that plagiarism is wrong in the context of essays, yet there is often a bit of a blind spot when it comes to presentations. However, making sure that you reference the sources of information in your presentation is not simply about avoiding plagiarism; just as importantly it is about showing the objectivity of what you are saying.

Precise

Scientific presentations should also be precise. As was highlighted in Chapter 7, 'Writing essays', the use of generalizations often clouds the meaning of what is written or indicates a lack of confidence on the part of the presenter. For example, what does the following phrase actually mean?

> The animals are fairly active at night, moving around quite a bit . . .

The phrases 'fairly active' and 'moving around quite a bit' have no real meaning as their interpretation depends entirely on the listener. Relative terms such as 'quite' or 'fairly', or phrases such as 'In general terms . . .' lack precision and should, wherever possible, be replaced with more precise phrases:

> During the hours of darkness, the animals forage widely, travelling up to 2 km from the nest site . . .

Clear

The final characteristic of scientific presentations that we have identified is that they should be clear. This does not just apply to simple, uncomplicated subjects. In fact, one of the marks of a good presentation and a good presenter is the ability to explain complex things in relatively simple terms. Using simple terms should not mean that the presentation is simplistic, but it does mean that the content needs to be explained in such a way that an average member of the audience can understand it without too much difficulty.

10.3 Analyse the question or brief

We have already identified that relevance is a key characteristic of effective presentations. We observed that a presentation needs to be both relevant to the question or brief you have been set, and relevant to the audience you will present to. A useful technique for analysing the question or brief is simply to write it in the middle of a piece of paper and then annotate it, as can be seen in Figure 10.2.

This technique enables you to begin to get your thoughts down on paper quickly whilst encouraging you to link your thoughts to the task you have been set, thereby helping you to keep your ideas relevant. It also allows you to make connections between ideas and to see, at a glance, how the presentation might be structured.

Figure 10.2 Analysing the question/title

This is how long it needs to be

Explain what it is

'Prepare and deliver a 5-minute presentation on the use of global positioning systems in the study of migrating populations. The presentations will take place at 12 noon on Friday 1st March in room 101.'

Look at lecture notes on this

This is when and where it is

Think of examples
- Wildebeest?
- Canada geese?
- Others?

Also, see chapter in core text

Try this—Analysing a presentation task

Consider the following task and produce an analysis that you could use to begin preparing for the presentation.

'Your next task is to give a five-minute presentation on recent developments in drug development for treating Alzheimer's disease. You will deliver the presentation in next Thursday's workshop in Lecture Room D.'

10.4 Research the topic

As we saw in Chapter 7, 'Writing essays', researching the topic can be a very time-consuming task. There are three things you can do to reduce, or at least control, the amount of time the research part takes.

1. *Analyse the brief first*, so that you have some clear ideas of what you are looking for before you start looking for it. It is easy to waste a lot of time researching material that you later discover is not particularly relevant; analysing the brief provides a focus for your research.

2. *Set a time limit for your research.* If you don't set a time limit, the research element can just go on and on, especially if you are being particularly diligent or are very interested in the topic.

3. *Look at source material in an appropriate order.* If you start off with an internet search it can take you off at all sorts of tangents. If, however, you start with lecture notes and key chapters from core texts, it provides focus for your research, so it is easier to discriminate when you get to the other sources (see Chapter 4, 'Working with different information sources'). A suggested order is as follows:

   - *start with your lecture notes*—check that you understand the information;
   - *identify the essential reading* from the relevant chapters in the core texts;

- when you have mastered the basics, *extend your reading* by looking at books that address a particular aspect of your topic;
- *look at recent research articles* to bring you up to date on the subject and to add the necessary level of detail.

It is important to note that jumping to internet sources too early can actually hinder your research. You need first to establish a clear focus that is relevant to the question or brief so that you can be highly selective if you do choose to supplement your research with internet sources.

10.5 Plan and order the material

During the research stage you will have probably gathered more material than you actually need, even if you have been quite disciplined in your approach. The temptation is to include too much material because you have spent the time finding it; however, this will often result in a presentation that is difficult to follow, unfocused, or too long (or all three!). The following stages will help you to decide what material you should include.

10.5.1 Choose your main points

The starting point for selecting and organizing your material should be to choose your main points. You might not be able to list all of the main points straight away, but it is important nonetheless to make a good attempt. As you continue in your preparation you may well come back to your main points to refine or reorganize them, but you need a basic framework at an early stage to focus your thoughts and provide a structure for the rest of your material. An example structure for a five-minute oral presentation on the use of GPS in the study of migrating populations was given earlier, as follows:

- what GPS is;
- applications to migrating populations;
- examples: wildebeest and Canada geese;
- conclusion.

This is a very sketchy outline, but it is enough information to get you started. A useful technique to help you draft your main points is simply to put your research notes to one side, then try to write your main points without looking at them, expressing each point in a few words or a short sentence. Alternatively, you could try to articulate your main points to a friend, because expressing ideas out loud can be a good way of helping you decide whether or not they are making sense. It may also be helpful to add some notes to your outline to give you an idea of what the focus of your presentation should be. For instance, from the list already given, it looks like the four bullet points (or at least the first three) are equally sized sections, which probably isn't your intention. It might be better, therefore, to annotate your main points with a bit more information, as follows:

- *What GPS is*—just brief background, enough for people to understand the basics.
- *Applications to migrating populations*—how the technology has been used in this context, and an overview of its applicability and value.

- *Examples*: wildebeest and Canada geese (more detailed descriptions of the application to these specific examples).
- *Conclusion*—summarize the main points and invite questions.

This is also the point at which you would begin to organize your main points into a logical sequence, so that each point links to and builds on the previous one. The sequence of your main points will be determined not only by what is logical, but also by what you are aiming to achieve. For example, if you are trying to build an argument you will want to move from background information to precise points of detail, or if you are explaining a process you might begin by explaining its purpose and then take your audience through each stage of the process step by step. So our oral presentation on migration might go something like this:

> What is GPS? GPS stands for Global Positioning System and most of you will be familiar with its application in the context of in-car navigation or the mapping software on your phone. As we can see on the second slide, the system is based on a network of satellites linked to ground stations around the world. To find where you are, the system measures the distances between your personal unit and three or four satellites, as you can see here . . . . The exact location is then worked out by triangulation.

What you need to keep in mind is that you are trying to communicate information in manageable chunks, helping the audience to understand the progression of your argument or process. It doesn't matter at this early stage if you can't sequence your main points precisely (for instance, will it be better to deal with wildebeest or Canada geese first?), because such a level of precision is not necessary at this point. The important thing is simply to choose what your main points will be and make a good attempt at the sequencing; you can always revise it later on.

10.5.2 Select supporting information

We have already identified that there is a temptation to cram too much information into a presentation, simply because you have found lots of information at the research stage. An important arbiter that we have referred to twice previously in this chapter is the relevance of the material (to both the brief and the audience). However, there are other criteria that are helpful to consider explicitly when you get to the stage of selecting supporting information. The other criteria that we will consider are clarity, authority, and interest.

Clarity

When deciding whether information should be left in or taken out, ask yourself this question: 'Is this information that I'm thinking of including going to make what I am trying to say any clearer?' Of course, this relies on you knowing what it is you are trying to say, which is why establishing the main points first is important. For instance, in your research, you may have found out lots of information about the GPS, including that the first experimental satellite for GPS was launched in 1978, but will this piece of information help to clarify what you are trying to say? If you have decided that the focus of your presentation is the examples themselves (wildebeest and Canada geese), then you will probably decide that incidental detail about the history of GPS is unnecessary.

Authority

Ask yourself: 'Is this information that I'm thinking of including going to make what I am trying to say any more authoritative?' The authoritativeness of information you use will depend largely on the source of that information; whether or not it is well researched, impartial, and up to date. This is usually easier to establish for publications such as textbooks and articles published in research journals than for web resources, especially freely editable sources such as Wikipedia. The authoritativeness of sources is addressed in Chapter 4, 'Working with different information sources'.

Interest

Finally: interest. Ask yourself this question; 'Is this information that I'm thinking of including going to make what I am trying to say any more interesting?' Perhaps you have found a picture, chart, or even a video clip, which will help illustrate a point. Perhaps you could bring in a GPS device to illustrate how compact they are and how easily the technology can be applied to the study of migrating populations. Consider anything that will bring what you are saying to life.

So, establishing whether information you have found will make your presentation clearer, more authoritative, or more interesting is a helpful means of deciding whether or not it should be included. However, we always come back to the bigger, more fundamental question, 'Is this relevant?' Also remember that when you decide to leave information out, it doesn't necessarily mean that the information is wasted: the information you leave out may have been helpful in leading you to other information. It may also be useful supplementary information if you are asked questions about your presentation, or it may be useful if you subsequently have to write an associated essay or report.

Try this—Adding interest to a presentation

Find three things that you could use in the presentation of migrating populations to add interest. These could be interesting examples of applications, pictures, video, etc.

10.6 Prepare your notes

Once you have analysed the brief, researched the topic, and planned and ordered your material, you are ready to prepare your notes. You have two main options available to you here: you could write them out word-for-word (known as *verbatim*) or you could use outline notes (e.g. bullet points). Alternatively, you could use a combination of these two types. We will consider each of these two options in turn and then look at how they can be combined to best effect.

10.6.1 **Verbatim notes**

Verbatim notes have many advantages and disadvantages.

Advantages of verbatim notes

Verbatim notes have the obvious advantage that you start the presentation knowing exactly what you intend to say. Knowing exactly what you intend to say can be reassuring and so boost your confidence. Verbatim notes can be particularly helpful, therefore, if you are lacking confidence in a particular presentation:

- perhaps because your first language isn't English;
- or you are presenting information that you are relatively unfamiliar with;
- or you are presenting complex material that requires a level of precision that would be difficult to achieve from outline notes;
- or you have some other reason for feeling nervous.

Disadvantages of verbatim notes

Verbatim notes do have significant disadvantages. There is a danger that you will simply read out your presentation and therefore won't communicate it effectively. When you look up towards the audience you might lose your place in your notes. Also, a read-out presentation is difficult to bring 'alive' because:

- you will lose eye contact with your audience;
- your voice will probably become more monotonous;
- you will have much less energy and enthusiasm about the delivery;
- as a result, your presentation will be less engaging.

How to use verbatim notes

Verbatim notes seem to have got themselves a bad name. Often people will tell you not to use them because of the very real disadvantages listed previously, but this advice usually comes from people who have become experienced presenters and so have forgotten about much of the anxiety that is often associated with presenting. However, as can be seen from the advantages, there are very real benefits, too. The trick is to use verbatim notes in the right way. We will take each of the potential disadvantages in turn and suggest a strategy to help you avoid that particular pitfall.

There is a potential danger that you will simply read out your presentation

Don't! Reading out the presentation will nearly always result in the problems listed. You don't need to memorize your notes, but you do need to have practised them sufficiently so that you can work from them as a prompt, as opposed to as a script.

Figure 10.3 Unhelpfully formatted verbatim notes

The use of GPS in the study of migrating populations

Hi, my name's Sam and my presentation is about how GPS helps scientists to study migrating populations.

This presentation is divided into three main sections, as shown on the first slide. First, I'll give a brief explanation of what GPS is. Then I'll explain how the technology has been applied to the study of migrating populations. Finally, I'll tell you about two specific examples: wildebeest and Canada geese.

First then, an explanation of what is GPS? GPS stands for Global Positioning System and, as its name suggests, it forms a worldwide navigation system meaning you should never get lost again! As we can see on the second slide, the system is based on a network of satellites linked to ground stations around the world. To find where you are, the system measures the distances between your personal unit and three or four satellites, as you can see here . . . The exact location is then worked out by triangulation (see, you always knew that trigonometry you slogged through at school would come in handy).

The first GPS systems were developed in the late 1970s for the United States military. Since then the technology has developed considerably and in the 1980s became available for civilian use. GPS is now used for many civilian purposes, including: map making, land surveying, and of course, navigation. The particular use we are interested in, however, is in its application to the study of migrating populations of animals. Studying migrating populations using GPS has only been possible since the technology required to track animals has reduced in size. The necessary tracking tags are now small enough to track many different sorts of animal, not just large ones. Tags then relay information, via the GPS satellite system, to monitoring units, thus allowing scientists to know the precise coordinates of the animals they are studying. Such detailed information is invaluable for scientists, as previously they were simply not able to know about an animal's whereabouts in such detail.

If you look up from your notes you might lose your place

Quite possibly. So make sure you format your notes in such a way so as to help you not lose your place. For instance, increase the font size, increase the line spacing, use paragraph breaks, use bullet points, use annotations, and highlight key words. Figure 10.3 shows verbatim notes that would be very easy for you to lose your place with; Figure 10.4 shows more helpfully formatted verbatim notes that dramatically reduce the chances of you losing your place.

A read-out presentation is difficult to bring 'alive'

This is because you will lose eye contact with your audience, your voice will probably become more monotonous, you will have much less energy and enthusiasm about the delivery, and as a result, your presentation will be less engaging.

Again, most of this will be helped by practice. If you can get to the stage where you have practised your presentation sufficiently so that your notes are a prompt, rather than a script,

Figure 10.4 Helpfully formatted verbatim notes

The use of GPS in the study of migrating populations

Don't start until you're ready

Introduction

Hi, my name's Sam and my presentation is about how GPS helps scientists to study migrating populations.

This presentation is divided into three main sections, as shown on the first slide.

1. First, I'll give a brief explanation of what GPS is.
2. Then I'll explain how the technology has been applied to the study of migrating populations.
3. Finally, I'll tell you about two specific examples: wildebeest and Canada geese.

Outline slide

1 min. max.

What is GPS?

First then, an explanation of what GPS is? GPS stands for Global Positioning System and, as its name suggests, it forms a worldwide navigation system meaning you should never get lost again!

As we can see on the second slide, the system is based on a network of satellites linked to ground stations around the world. To find where you are, the system measures the distances between your personal unit and three or four satellites, as you can see here . . .

you should be able to think about the communication, as well as content issues—issues such as making eye contact with the audience, varying the tone of your voice appropriately, and generally trying to be more engaging.

10.6.2 Outline notes

Outline notes also have advantages and disadvantages.

Advantages of outline notes

Having your notes in only outline form can free you up to focus on engaging with the audience, rather than focusing on your notes. Also, if you are not reading your notes out, your voice and manner will probably be more natural, which will also help engage your audience. There is no temptation to read your notes out verbatim, which, as we highlighted earlier, causes problems.

Disadvantages of outline notes

There are two main disadvantages of outline notes. First, you might be more nervous about your presentation because you have fewer notes to rely on. Secondly, you may find that your mind 'goes blank' during your presentation because your outline notes aren't a sufficient prompt to remind you of what you intended to say.

How to use outline notes

As we have seen with verbatim notes, outline notes have both advantages and disadvantages; the trick is to use outline notes in the right way. The only way around these potential disadvantages is practice: you need to rehearse your presentation a sufficient number of times so that you know your outline notes are full enough to remind you of what you need to say. Of course, while you are testing the adequacy of your notes in this way you will also be learning the material. Remember though, just because you can get your presentation right once or twice in the privacy of your bedroom doesn't mean that everything will be fine in the more pressured situation of a tutorial class or assessed presentation.

10.6.3 Combine note types to maximize benefits

Given the potential disadvantages of each notes technique, probably the best method to employ is a mixture of the two. Where you are confident that you only need short bullet points to prompt you, you can use outline notes, for instance, for the introduction, as shown in Figure 10.5. However, where the material is more complex and precision is more important, you can use at least fuller, if not verbatim, notes, as shown in Figure 10.4.

10.6.4 Openers and closers

In addition to planning the introduction, main points, and conclusion of your presentation, it is also important to think about how you will open the presentation and how you will close it. It is a good idea, regardless of what type of notes you choose to use, to know exactly what your opening statement is going to be (the 'opener') and exactly what your closing statement is going to be (the 'closer'). If you have a faltering start, perhaps tripping over your opening words or not saying them clearly, it is likely that you will lose confidence

Figure 10.5 Outline notes for the introduction

and composure, which can be very difficult to regain. Opening statements need only be simple sentences such as:

> 'Hi, my name's Sam and my presentation is about how GPS helps scientists to study migrating populations.'
> 'Hi, my name is Alex and I've got five minutes to talk to you about . . .'

Simple statements such as these get you off to a positive start and make it clear to the audience what you are doing. Equally, closing statements are important, too. Too often, presentations finish with a mumbled apology: 'Err, that's it, I've finished . . .', then the presenter sits down. A much better way to finish is a clear conclusion followed by a closing statement such as:

> 'So, in this presentation we have seen how GPS can be used as a very effective tool for monitoring the migration of very different types of animal, but also that there are some specific limitations that need to be addressed.
> Thank you very much for your attention and I will be happy to try to answer any questions.'

Try this—Preparing a presentation

Choose a topic in biosciences that you are interested in or use a topic you have been given on your course. Prepare slides sufficient for a five-minute presentation. Ask a friend to look through them and provide you with one good point and one point for improvement. If this is a class assignment, you should provide the same feedback on your friend's slides.

10.6.5 A note about working from visual aids

An alternative to speaking from either verbatim or outline notes is to work from a visual aid. This is when you use a visual aid (e.g. a PowerPoint slide or an object) to prompt what you are going to say, rather than your notes themselves. This can work well because it frees you from your notes entirely, yet can still give you sufficient prompts to remind you of what you want to say. However, as with verbatim and outline notes, it all depends on how the method is used. The method will work badly if your slides are essentially copies of your verbatim notes (as shown in Figure 10.6) and you end up reading them out. However, if you either condense the bullet points to key words or phrases, or better, use an image to illustrate what you want to say (as shown in Figure 10.7), you will find it much easier to talk from (this will be covered in more detail in Chapter 11, 'Delivering scientific presentations'). As a tip, if you do have text on your slides, avoid reading it out: if they are bullet points, expand around them; if you have used fuller text, for example a quotation, then allow the audience to read it, rather than reading it aloud for them.

Figure 10.6 A difficult-to-work-from slide

> What is GPS?
> - The global positioning system (GPS) is the only fully operational global navigation satellite system
> - More than 25 GPS satellites are in orbit around the Earth, transmitting signals that are picked up by GPS receivers, which, in turn, establish the receiver's location, speed and direction
> - GPS is a crucial aid to navigation

Figure 10.7 An easier-to-work-from slide

10.7 Practice

Finally, in this section on preparing for presentations we need to highlight the importance of practice. How effective a presentation is will depend to a large extent on how much you have practised it. Practising a presentation is important because if you know your presentation well it frees you up to think about the communication, as well as content issues (style, as well as substance).

We noted previously in Section 10.6.2 'Outline notes', 'How to use outline notes', that just because you can get your presentation right once or twice in the privacy of your bedroom, it doesn't mean that everything will be fine in front of an audience. However, the value of practising in your own environment is not to be underestimated, and there are several ways you can make it more effective.

10.7.1 Practise against the clock

It can be very difficult to judge how much time your presentation will take without actually practising it out loud and against the clock. Depending on the style of notes you have used, sometimes you will look at your notes and think that it will barely last two minutes, never mind five. Other times you will think you have got far too much material, but in reality you will get through it far quicker than you thought. The only way to have a reasonably accurate idea of how long your presentation will take is to practise it out loud. Of course, by doing this you are not only getting an idea of how long it takes, you are also learning the material and refining its delivery.

10.7.2 Practise in front of a mirror

Just as it is difficult to know how long the presentation will take, it is also difficult to know how you will look. The only way to get an idea of this is to practise the presentation in front of a mirror or, better still, record yourself on your phone. You will probably feel a bit self-conscious doing this, but it will give you some very useful feedback about how you are presenting yourself, as well as how you are presenting your material. It is a particularly helpful way of assessing whether or not you are using your body language appropriately (see Section 11.3.3, 'Think about your body language').

10.7.3 Practise in front of your friends

Most usefully, you could practise in front of some willing friends. This could give you feedback on your presentation's length, your body language, the structure, in fact all aspects of your presentation. It can be a bit embarrassing and you will doubtless make mistakes, but as long as your friends are going to give some constructive comments, and you can deal with the potential embarrassment, it will give you invaluable feedback and so improve your final presentation, and thereby its grade, significantly.

 Chapter summary

The ability to present information verbally to groups of people is an increasingly important skill. It is of particular relevance to scientists, as it is an important means of communicating research and it is also an important transferable skill that employers value highly. This chapter began by identifying the characteristics of effective presentations and then identified the stages necessary to prepare presentations. Analysing the question or brief, researching the topic, planning and ordering the material, selecting supporting information, and preparing your notes, were all identified as important aspects. Finally, the chapter addressed the issue of practising your presentation to ensure that you deliver it as well as you possibly can. This brings us on to Chapter 11, 'Delivering scientific presentations'.

11 Delivering scientific presentations

 Introduction

You may be wondering why we have devoted an entire chapter just to the delivery of scientific presentations. Chapter 10, 'Preparing scientific presentations', dealt in some detail with the preparation aspects of scientific presentations, so is a whole chapter purely on the delivery aspects really necessary? We think it is. As we identified in Chapter 10, 'Preparing scientific presentations', presentations can be divided into two overlapping, but nonetheless distinct areas: content, and the communication of the content (or 'substance' and 'style'). We have said that both of these elements are important and you need to be careful not to focus on one at the expense of the other. Just because you have prepared a good presentation (the content) it doesn't mean that you will be able to deliver it in a competent fashion (the communication of the content). Communicating the content involves a very different set of skills to preparing the content, so this chapter will look at how you can communicate your content to best effect.

In this chapter we will begin by highlighting common concerns people have about presentations and how to address them, then we will think about how to use visual aids effectively, and finally some key techniques to help your presentations run smoothly. First, let's think about some important differences between written and spoken presentations.

Written versus spoken presentations

A useful starting point when considering the delivery of a presentation that you have prepared is to think about the similarities and differences between written and spoken presentations. One of the reasons some students get anxious about giving spoken presentations is that they consider them to be very different from written presentations. Of course, in many respects, the two forms are different, and we will come to the differences shortly, but there are many similarities, too. In terms of the process that you need to go through in order to create written or spoken presentations, both forms require you to:

- analyse the question or brief;
- research the topic;
- plan and order the material;
- write the text or notes;
- review and redraft the text or notes.

In addition to these, in an academic context, both forms require you to present and analyse evidence, structure your argument, and reference your sources. When it comes actually to delivering the presentation, however, the major difference between the two forms is that, for a written presentation, all that is required is for you to hand it in (whether that be on paper or via email or web submission), whereas delivering a spoken presentation requires a lot more effort.

There are other important differences between written and spoken presentations. For example, when you are reading a textbook or journal article, and you come across a complex argument that you don't understand you can skip back and re-read the argument until you do understand it. Or if you come across a large quantity of data that is difficult to take in, you can study the data more closely, or even look up additional information in other sources, to make them easier to understand. When taking in information in the context of a spoken presentation, however, it is often much more difficult to assimilate complex arguments or large quantities of information simply because you do not have this ability to skip back as you would with written material. Equally, if, when you are reading, you find yourself getting tired and your attention is wavering, you can simply stop reading, have a break, and come back to it later when you are feeling more alert. Clearly, when listening to a presentation, apart from walking out and choosing to miss the rest of the presentation (which would be a bit rude), you don't have this option. So in a presentation, presenters have to work hard to make sure their listeners stay with them, in terms of both their understanding and their attention. This chapter will explain how you can do this.

11.1 Common concerns and how to address them

The differences between written and spoken presentations are often the cause of concern or anxiety about delivering a presentation. For example:

- What if I'm so nervous I forget what I was going to say?
- What if the technology doesn't work?
- What if I lose my place in my notes?
- What if the audience doesn't seem to be following what I'm saying?
- What if I get asked questions that I don't know the answers to?

These, and other concerns, are important things to consider. While listing possible concerns in this way can seem a little overwhelming, it is important to identify them so that something can be done about them. You might be able to relate to all of these concerns, or perhaps to just a few of them, but it is likely that you will experience all of them at some point or other, so it will be useful to address each of them in turn.

11.1.1 Deal with anxiety

It's all very well saying 'deal with anxiety'; it is much more difficult to actually do it. It is also quite difficult to write about how to deal with anxiety, because different things make different people anxious. However, for the purposes of this chapter we are going to assume that standing up in front of a group of people to give a presentation is something that causes many people to be anxious or nervous. If it doesn't, it probably should!

Table 11.1 Replacing negative thoughts with positive ones

Negative thought	Positive replacement
'It will be a disaster.'	'I will aim to do the best I can.'
'I never do any good at this kind of thing, it's bound to go horribly wrong.'	'Just because I had a problem with this is in the past does not mean that things are bound to go wrong.'
'I will fail my degree and never get the career of my choice if I don't do well in this presentation.'	'The marks for this presentation are only a small percentage of my overall degree. If I don't do as well as I would like there will be other opportunities to improve my marks.'

We have deliberately titled this section 'deal with anxiety', rather than, for example, 'get rid of anxiety' for a couple of reasons. First, you can't get rid of all your anxiety, so it would be unrealistic to try; and secondly, anxiety or nervousness can be helpful, so even if you could, you don't want to get rid of it altogether. Anxiety can motivate you to do the work needed, so that you are well prepared for the presentation, and it can also make you more alert and energized during the presentation. However, if your anxiety becomes too pronounced, difficulties can occur, which may impair your ability to prepare effectively for, and perform during, the presentation.

A useful strategy in dealing with anxiety is to identify what your concerns are so that you can address them, which is what this section is about. In general terms, however, the following strategies can be helpful:

- try to replace negative thoughts with positive ones (see Table 11.1);
- practise steady, deep breathing before and during your presentation;
- have a bottle of water to hand—it will help stop your throat drying out, but better still, if your mind goes blank you can simply stop and take a sip, which usually gives you just enough time to gather your thoughts and regain your composure without the audience noticing.

Try this—A one-minute presentation

Get together with a small group of friends. Each choose a topic with which you are comfortable, for example, your favourite hobby, holiday destination, or similar topic with which you are familiar. Now prepare a very short presentation on this topic. Each of you should now present the talk to the rest of the group. The talk should be no longer than one minute. One member of the group should keep time.

11.1.2 Have a back-up plan

The second concern that we identified was 'What if the technology doesn't work?' This is a legitimate concern and one that requires consideration. Broadly speaking there are two possible causes of technology apparently not working: first, it could be a genuine fault with the technology; or secondly (and perhaps more commonly), it could be a lack of competence on the part of the person operating it. Both types of problem can be minimized by having a back-up plan, but be careful not to jump too quickly to using your back-up plan, when it could be, for instance, that you are just pressing the wrong button!

It is important, therefore, that you familiarize yourself with any technology that you intend to use. Amid the pressure and stress of a presentation about to start, or already started, it can be easy to panic and not think straight. Make sure you understand how to use what you are using. If possible, practise not just with something like it, but with the *actual* technology in the *actual* room where you will *actually* be presenting. The closer you can get to practising with the real thing, the more confident and prepared you will be when it comes to the presentation.

One common source of technological problems is the transfer of the electronic file to the computer being used to deliver the presentation, for example, because of failure of a memory stick. If you have the chance to access the venue before your talk, it is always a good idea to copy the talk across onto the desktop of the computer and check it is working. If you can't do that, then having the talk on a back-up medium is also a good idea; for example, have a spare memory stick, or have the talk emailed to yourself so you can access that way.

For those occasions when the problem is with the technology, rather than the person operating it, this is where you will have to use your back-up plan. Some forms of technology are less reliable than others, and the more multi faceted the output the more potential there is for it to go wrong. For instance, if you are using slides, video, audio, and a live internet connection, there are a lot of things that can potentially go wrong! You need to consider whether or not such a range of technology is necessary to your presentation. There would be much less risk if you used less technology and it would be much simpler. Indeed, it may even improve your presentation (more technology does not necessarily mean better presentations).

11.1.3 Make your notes work for you

The third concern that we identified was 'What if I lose my place in my notes?' As we pointed out in Chapter 10, 'Preparing scientific presentations', there is a lot you can do to help yourself here. First, make sure you use an appropriate type of notes: outline or verbatim (or a combination of both). Secondly, make sure you format them in a way that makes them easy to use and so it is less likely that you will either lose your place (a possible problem with verbatim notes) or forget what you were going to say (a possible problem with outline notes). Chapter 10 also highlighted the benefits of combining these note types to maximize the benefits, using outline notes where you don't need much prompting and verbatim notes where you do.

11.1.4 Help the audience follow what you are saying

The fourth concern was 'What if the audience doesn't seem to be following what I'm saying?' This is clearly an important consideration for the benefit of the audience, but it is also a significant factor in your own performance as a presenter. If you begin to notice, during your presentation, that your audience is distracted and perhaps seemingly confused, it can be very difficult not to let that influence your confidence and you can begin to lose your composure as a result. Obviously, it is important not to be over-sensitive here; it can be easy to interpret normal audience behaviour, such as someone yawning, as a signal that your presentation is going really badly, when this probably is not the case. It is likely that there will always be some people in the audience who have had a bad night's sleep, or are thinking about what they are going to do at the weekend, or perhaps just exhibiting body language and facial expressions that are out of sync with their genuine response. Don't let one or two negative responses unduly influence your confidence. Instead, there are several things you can do during your presentation to help your audience follow what you're saying:

- use verbal signposts;
- use your voice;
- be aware of your audience's needs.

Let's consider each of these in turn.

Use verbal signposts

In Chapter 10, 'Preparing scientific presentations', we identified that one of the character-istics of effective presentations was that they should be well structured. If a presentation is well structured, then the audience can see the direction the presentation is taking and so should be able to follow it more easily. We said that one of the ways you can do this is by letting the audience know what the structure will be; both at the beginning and as you go along.

For example, at the beginning of a presentation you should map out where you are going by telling the audience what you will cover. The example presentation brief that we intro-duced in Chapter 10 was:

> Prepare and deliver a five-minute presentation on the use of global positioning systems (GPS) in the study of migrating populations.

For this presentation, letting the audience know what the structure will be may involve you saying something like this (as seen in Chapter 10): 'This presentation is divided into three main sections. First, I'll give a brief explanation of what GPS is. Then I'll explain how the technology has been applied to the study of migrating populations. Finally I'll tell you about two specific examples: wildebeest and Canada geese.'

In addition to mapping out the structure at the beginning of the presentation, it is also helpful to give the audience prompts as you go along. If you were reading a written presenta-tion you could simply flick through the pages, and read the headings and subheadings, but clearly you can't do this with a spoken presentation. So help your audience follow what you are saying by telling them what the structure will be and then use pointers as you go along to make it clear where you are up to. The examples from Chapter 10 were:

- a single word such as 'First', 'Secondly', and so on;
- linking statements such as 'So that's the background to the technology, let's now think about its application . . .';
- concluding statements such as 'In conclusion . . .' or 'Finally . . .'.

Use your voice

Use your voice to help your audience follow what you are saying. You can use it in many different ways by varying the volume, pace, and pitch.

Volume

Make sure that your voice is loud enough for your audience to hear clearly (it's surprising how many presenters don't do this—nervousness will often make you speak more quietly). Speaking too quietly (or too loudly) can make it difficult for your audience to follow your presentation. In normal conversation people tend to raise or lower their volume for emphasis. For example, a person may speak loudly when giving an instruction, but softly when apologizing. Try to use these normal conversational variations in your own presentation (but don't go too quietly!)

Pace

Make sure that the speed of your delivery is easy to follow. If you speak too quickly (again, this is often caused by nervousness) or too slowly your audience will have difficulty in following your talk. To add life to your presentation, try changing the pace of your delivery. A slightly faster section might convey enthusiasm. A slightly slower one might add emphasis or caution. As you become more experienced this kind of variation will happen much more naturally, as it does in conversation (especially as you become better at dealing with the anxiety), but it's a good idea to be deliberate about it while it doesn't come naturally to you.

Pitch

The pitch of your voice also varies in day-to-day conversation and it is important to play on this when making a presentation. For example, your pitch will rise when asking a question; it will lower when you wish to sound severe.

Be aware of your audience's needs

If you are giving a five- or ten-minute presentation to a group of your peers and a tutor, then you will be limited as to how much you can actually do about your audience's needs. You will probably be on quite a tight schedule (often you will lose marks for running over time) and if there are several of you presenting, then it will often be a case of just making sure the presentations fit into the allotted session. Additionally, when you are relatively new to presenting, you will want to concentrate on just delivering what you have prepared, rather than making any changes to your presentation as you go along in response to the reaction you are receiving. However, if you are at least half way down a list of, say, six presentations that are scheduled for a session, there is one simple thing you can do (if your tutor doesn't suggest it)—simply suggest a brief break. The break does not have to be long and it will probably be best if people don't even leave the room. However, if you suggest a break of a minute or so, while you are setting up, perhaps even suggesting that people stand up and walk around a bit, it should be just enough for people to wake up and re-energize themselves so that when you start your presentation they are that bit more attentive, making it easier for both you and them.

11.1.5 Respond to questions appropriately

The final concern we identified was 'What if I get asked questions that I don't know the answers to?' This is a common concern, because it is the bit of the presentation that you have least control over and it can be difficult to know what kind of questions you might be asked. If you are doing an assessed presentation, one important piece of information to find out is whether or not you are required to invite questions, and if so, whether or not your responses count towards the overall mark you receive. Whether or not the questions (or more accurately your answers!) count towards your mark, there are several things you can do to make this part easier.

Create a list of potential questions and prepare answers

We said earlier that it can be difficult to know what kind of questions you might be asked, but it is not impossible. In fact, with just a bit of thought you can probably come up with several potential questions that you might get asked. Questions will usually focus on areas of the presentation that the audience found particularly interesting, or perhaps a section that needed more explanation or additional context. Obviously, you can't predict all the questions you might be asked, but you will be able to predict some, and that will help to take some of the pressure off.

> #### Try this—Identifying possible questions
>
> Show a draft of your slides and presentation notes to a friend or peer. Ask them to come up with two or three questions that occur to them while they are reading your material. Draft an answer to each question and get feedback on the helpfulness of your answer from your colleague.

Say whether you will be taking questions and when

It is a good idea to state clearly at the beginning of your presentation whether you will be taking questions and, if so, when. That way the audience is clear about if and when they can ask questions, but also it means that you are more in control of the situation, which is important—lack of control, as identified earlier, can make the questions part of a presentation difficult. Sometimes people like to deal with questions as they go along as it can allow more interaction. However, it is usually easier to deal with questions at the end so you can deliver your talk free from interruptions, which you might find distracting. Taking questions at the end also means that you keep control of the timing of the talk, which you may find difficult to do if there are questions being asked as you go along.

Ask for clarification

When you get asked a question it is often unclear what the question actually means. If you are going to answer the question that the questioner wants to ask, rather than the one you think they might be asking, then it is important to ask for clarification. For instance, you might get asked the question:

> Do all the animals follow the same migrating pattern year after year?

A good way of clarifying the question is to rephrase it and then ask if that is what was meant. For example:

> When you say 'all the animals' do you mean you mean all migrating populations of wildebeest or all migrating populations of Canada geese?

Clarifying questions in this way is useful for a number of reasons:

- it gives the questioner a chance to refine the question;
- it ensures that the rest of the audience has heard and understood the question;
- it enables you to check that you have understood the question;
- it also gives you thinking time to prepare a better answer.

Answer the question

Preparing potential questions and answers, saying if and when you will take questions, and asking for clarification will all help you to answer questions more appropriately. However, when you actually open your mouth to answer the question make sure you:

- *answer the question you have been asked*, rather than the one you want to answer (seeking clarification helps here);
- *are focused and to the point*—answer the question and then stop, don't start waffling;
- *are honest about the limitations of your knowledge*—say if you don't know the answer (it will usually be obvious if you are bluffing);
- *say if a question is beyond the scope of the presentation*—e.g. 'That's a good question but my research for this presentation focused on tracking migrating populations by GPS, not by any other means').

11.2 Use visual aids effectively

Using visual aids is relatively easy; it's using them appropriately that can be difficult. You will be limited to a certain extent by what is available, but it is equally important to consider what is appropriate; just because you *could* use something doesn't mean that you *should* use something. The list in Section 11.2.1, 'Different types of visual aids', shows the variety of visual aids available to you, and it is important that you recognize this variety;

choosing the right form of visual aid is the first step in using them appropriately. Too often, presenters default to using PowerPoint (or equivalent) slides, without considering whether or not this is appropriate. For instance, a digital projector and laptop may be available to you, but using on-screen presentation software may be over the top if you are simply required to give an informal update regarding a project to a small tutorial group. In this situation, a concise handout identifying key points that you can chat through might be more suitable.

11.2.1 Different types of visual aids

The more common types of visual aid are as follows:

- *Whiteboard*: a large white board, usually fixed to the wall, can be written/drawn on with dry-wipe marker pens, either pre-prepared or spontaneously produced.
- *Flip chart*: a large pad of paper (A0), attached to a portable stand, can be written on with marker pens, either pre-prepared or spontaneously produced.
- *Handouts*: usually A4 paper, handed out at an appropriate point (or points) during the presentation.
- *Objects*: could be any physical object relevant to your presentation, shown from the front or handed round to your audience.
- *Video*: usually shown via a computer; there are various formats—DVD, digital media file, or streamed online (e.g. via YouTube).
- *Overhead projector slides*: hand-written or printed onto transparencies, either pre-prepared or spontaneously produced.
- *Presentation software (Microsoft PowerPoint, Google Slides, Apple Keynote)*: a linear sequence of slides produced on a computer and projected onto a screen via a digital projector; can be static or animated.
- *Prezi*: a non-linear form of presentation software that is produced on a computer (via https://prezi.com[1]) and projected onto a screen via a digital projector, but uses a canvas with frames that can be zoomed in and out of, rather than linear slides.

Each type of visual aid has advantages and disadvantages. These are summarized in Table 11.2.

11.2.2 When to use visual aids

Visual aids can be used throughout your presentation, but too many visual aids can become distracting. Try to restrict the use of visual aids, therefore, to key parts of your presentation. Some suggestions of when to use visual aids follows (we will address what visual aids to use shortly).

---

[1] Educational licenses are available if you sign up with your university email address, see http://prezi.com/pricing/edu/

Table 11.2 Advantages and disadvantages of various visual aids

	Advantages	Disadvantages	Suggestions
Whiteboard	• They are usually big—so lots of space to write/draw • They are particularly good for drawing processes or equations because there is lots of space • Good for things you want to leave up and refer to at different points during your presentation • Good for recording comments or questions from the audience as you go along • Can be pre-prepared or spontaneous	• If you write on it during your presentation you will be turned away from the audience as you are writing, which limits communication • If you run out of space you will need to rub off at least some of what you've previously written • You need to be able to write legibly and large enough for the audience to see	• There aren't always appropriate pens available—so take your own • Write large and clearly in dark colours (black or blue)
Flip chart	• Good for things you want to leave up and refer to at different points during your presentation • Good for recording comments or questions from the audience as you go along • Sheets can be torn off and stuck on the wall for reference—or flip back through the pad • Can be pre-prepared or spontaneous	• Usually smaller than a white board so not suitable for larger groups • If you write on it during your presentation you will be turned away from the audience as you are writing, which limits communication • You need to be able to write legibly and large enough for the audience to see	• There aren't always appropriate pens available—so take your own • Write large and clearly • Try to keep to one topic per sheet
Handouts	• The audience has something to take away for reference • Particularly good for supplementary information, e.g. references	• If handed out at the wrong time they can be distracting because the audience stops listening to the presenter and starts reading the handout • If handouts are very detailed it is more likely that the audience will listen less	• Use them for information that is too detailed to put on a slide • Think carefully about when and how to give them out

(Continued)

Table 11.2 Advantages and disadvantages of various visual aids (*Cont.*)

Objects	• Not a commonly used form of visual aid, so can have added impact • Other visual aids are mostly just looked at, objects give the audience something to touch, too, and are therefore more tangible and memorable	• If you have only one copy of the object it will take time to hand round (so maybe better to just show from the front) • Because of the interest they can generate it is important that the object is central to your presentation, otherwise it will just be a distraction	• Make sure the object is directly relevant to the content of your presentation
Video	• Can add interest and help grab attention	• Can be hard to find a directly relevant video • The technology can be a little unpredictable	• Make sure you keep the clip short • Check the technology works • If using presentation software, find out how to embed video in a slide (this can be done with YouTube videos, too)
Overhead projector slide	• Projected image is usually large enough for even large groups to see • Can be hand-written or printed • Can be good for showing the development of a process or building up a diagram	• You need to be able to write legibly and large enough for the audience to see • Transparencies are expensive	• Make sure text is large enough to be seen from the back of the room • Keep slides simple, clear, and concise
Presentation software	• Quick and easy to produce high-quality slides • Animation features can be helpful	• Some of the preset designs are unhelpfully formatted in terms of colour and background • Animation features can be very distracting if used inappropriately	• Keep slides simple, clear, and concise • Use animation only when appropriate • Black the screen when you're not using it (press 'B' when in slideshow view—going to the next slide will take it out of black mode)
Prezi	• Can help present ideas more intuitively—zooming in and out of aspects of ideas to show how they connect and interrelate	• As with other presentation software, can be used badly • Too much zooming in and out can make the audience feel dizzy • Can take a long time to prepare	• Use only when the content of your presentation lends itself to be displayed in this way • Using one of the templates can save time, but remember the content needs to lead the choice of the template, not the other way around

Introduction

You could use visual aids during the introduction to do some of the following:

- display the title of your presentation and perhaps your name;
- define particular terms or units you are going to use that your audience might not be familiar with;
- indicate what the structure of your presentation will be;
- display an image that introduces your subject or theme;
- highlight a question that you intend to answer during the course of your presentation.

Main points

You could use visual aids during the main body of your presentation to do some of the following;

- highlight your main points as they arise, using appropriate words or images;
- indicate the transition between the main points;
- summarize your results or data visually using graphs, tables, maps, etc.;
- display key evidence from your research to support your argument.

Conclusion

You could use visual aids during the conclusion to do some of the following:

- summarize your main points;
- present your conclusion in a succinct phrase or image;
- list your references to enable your audience to read more on the topic.

11.2.3 What visual aids to use

In Chapter 10, 'Preparing scientific presentations', we suggested a number of questions to help you select supporting information for a presentation. You can use these same questions to help you decide whether or not to use a particular visual aid. The fundamental issue is whether the material represented in the visual aid is relevant to the presentation, and if it is you, then need to ask yourself, 'Will this visual aid make my presentation . . .':

- clearer?
- more authoritative?
- more interesting?

If your visual aid doesn't do any of these three things, don't use it.

Try this—Choosing visual aids

For our sample presentation on GPS and migrating populations, a possible outline for the presentation was: 'Short explanation of what GPS is, how the technology has been applied to study migrating populations, and examples of wildebeest and Canada geese.' Imagine you were considering using the following visual aids for this presentation:

- a flip-chart page outlining the structure of your presentation (Figure 11.1);
- a handout giving supplementary details on how GPS works;
- pieces of card cut to 12 mm × 15 mm to illustrate the size of a GPS receiver;
- a video clip (2 minutes 41 seconds) of migrating wildebeest streamed from the National Geographic website;
- a hand-drawn overhead projector slide showing the migration path of Canada geese, as determined by GPS tracking studies;
- an image of a Canada goose on a PowerPoint slide (Figure 11.2);
- a Prezi showing the migration path of wildebeest, zooming in on the start and end points.

Based on the questions listed previously (Section 11.2.3, 'What visual aids to use'), decide whether or not you should include these visual aids in your presentation, and why.

Figure 11.1 Flip chart summarizing presentation content

OUTLINE

- What is GPS?
- Applications to migrating populations
- Examples
  - Wildebeest
  - Canada geese
- Conclusion

Figure 11.2 Image of a Canada goose in flight

Choosing visual aids feedback

We identified in Section 11.2.3, 'What visual aids to use', that the fundamental issue when choosing whether or not to use a visual aid is its relevance: is the material represented in the visual aid relevant to the presentation? If you decide it *is* relevant, you then need to question whether the visual aid makes your presentation clearer, or more authoritative, or more interesting. We will deal with each of the suggested visual aids in turn, testing them against each of these criteria.

A flip-chart page outlining the structure of your presentation

This is clearly relevant to your presentation because it summarizes its content (see Figure 11.1). Although it won't make the presentation more authoritative or more interesting, it could help make it clearer. If you were to use a flip chart in conjunction with another form of visual aid, for example an overhead projector, it would have the added advantage of the outline always being there for you to refer to. Flip charts are probably best used only when the audience is relatively small and close by.

A handout giving supplementary details on how GPS works

This is potentially relevant, but there is a danger of missing the main focus of the presentation. The presentation is primarily about the application of GPS in the study of migrating populations, it is not a presentation on GPS itself. A brief explanation of what GPS is and how it works is useful background information, but it only features in the presentation so that your audience can understand the main focus—its application to your two chosen migrating populations. So, basic information about GPS will help make your presentation clearer, but it is only there to provide context. It may add greater authority to the presentation, but this depends largely on the source of the information (see Chapter 4, 'Working with different information sources'). It could add interest, but this depends on how you communicate it (because of its potential lack of relevance it could actually make the presentation less interesting). However, if you genuinely think that the information is useful, a handout is probably the best format for it. Handouts are good for communicating (among other things) supplementary information; your audience can just refer to the handout if they are interested. Be careful about when you give them out though—you don't want your audience having their heads down reading when you are trying to move on to the next point!

Pieces of card cut to 12 mm × 15 mm to illustrate the size of a GPS receiver

This might sound a little bit unconventional, but that doesn't necessarily make it a bad idea. Again, it shouldn't be the main focus of the presentation, but it would illustrate well how the technology can be applied to the context in question. One of the reasons GPS can be used to track animals is because the tags required can be so small. Just saying that they are '12 mm × 15 mm' doesn't have much impact; showing with a visual aid how small they can be is much more striking. Better than saying '12 mm × 15 mm' would be to say 'they are about the size of your thumb nail', but actually handing out an appropriately sized piece of card adds an extra dimension, giving people something to hold and touch and think about. So using such a visual aid would make the point more clearly—it wouldn't add any authority, but it might add some interest.

A video clip (2 minutes 41 seconds) of migrating wildebeest streamed from the National Geographic website

This might sound exciting, but it is probably inappropriate for a number of reasons. It is potentially relevant in terms of the brief because it is about the migrating habits of one of your chosen populations, but it's probably not relevant in terms of the level. In Chapter 10, 'Preparing scientific presentations', we said that a presentation should be relevant not only with regard to the brief, but also with regard to the audience. A video designed for a popular audience is probably not going to be relevant for academic purposes. It probably won't make your presentation any clearer, unless it specifically addresses the use of GPS technology. It's unlikely to add authority because, even though it is from a credible source, the level, as identified before, is too popular. It might add interest, but the clip is far too long; if you were to use the entire clip it would take up more than half your allotted time for your presentation.

A hand-drawn overhead projector slide showing the migration path of Canada geese, as determined by GPS tracking studies

If you like the idea of using cutting-edge technology you might be reluctant to use a hand-drawn overhead projector slide. However, let's think about our criteria. Such a slide would be relevant because it would be illustrating a very important part of your presentation. It would make the content clearer by representing visually the information that GPS technology has been used to gather on this particular migrating population. It adds authority because the GPS tracking studies are a primary source of information. It might even add interest because you are illustrating the vast distances these birds fly and the places to which they travel.

An image of a Canada goose on a PowerPoint slide

The image is relevant in terms of the brief because it is a sample of one of your chosen populations. However, by itself it is not communicating very much information. It could be argued that it will help make your presentation clearer (because it makes it obvious which bit of your presentation you are on), and that it doesn't add any authority, but it could add some interest. It is appropriate to use images such as this (Figure 11.2), but make sure you have a clear purpose for using them (and as long as you acknowledge the source, see Chapter 5, 'Academic integrity and avoiding plagiarism'). Don't just think, 'there's a picture of a Canada goose—I'll use that'; ask yourself why you need to use it. Your audience probably knows what a Canada goose looks like, so that's not your reason, but your reason could be to use it as a background to the start of your section on Canada geese so that the audience are clear that you have moved on from wildebeest.

A Prezi showing the migration path of wildebeest, zooming in on the start and end points

This is the more hi-tech version of the hand-drawn overhead projector slide, but with the added benefit that you can show movement; using a map as a background image and zooming into the start and end points. As with the overhead projector slide it would be relevant because it would be illustrating a very important part of your presentation. It would make the content clearer by representing visually the information that GPS technology has been used to gather on this particular migrating population. It adds authority because the GPS tracking studies are a primary source of information. It is certainly an interesting way to present the information because you are illustrating the vast distances these animals move and the places to which they travel.

11.3 Master some key techniques

We have addressed some common concerns that people often have about giving presentations and have considered how to use visual aids appropriately. Finally, it's important to think about some key techniques that you need to develop if you are going to deliver your presentation effectively.

11.3.1 Relax!

The first key technique is to try to relax. Clearly, this is easy to say, but much more difficult actually to do. However, if you can relax, even just a little bit, it will help your presentation a great deal. Presenters who are tense will deliver their presentation in a stilted, unnatural, and nervous manner, whereas a presenter who is relaxed (or slightly less tense at least) will be more natural and more confident in their delivery. We addressed many of the issues that will help you to be more relaxed in Section 11.1.1, 'Deal with anxiety', but we also wanted to highlight it here as a key technique because it makes such a positive difference to a presentation.

11.3.2 Be conversational

In Chapter 10, 'Preparing scientific presentations', we highlighted the importance of getting to the stage where you have practised your presentation sufficiently so that your notes (whatever their format) act as a prompt, rather than as a script, thus freeing you up to think about the communication, as well as content issues. Something to aim for is to try to be conversational. Being conversational doesn't necessarily mean being informal or chatty (this is often not appropriate in an academic context), but it does mean that you are speaking in a natural way, as if you were actually having a conversation. Clearly, a five-minute presentation is a fairly one-way conversation (!), but if you can aim to express yourself in normal, everyday (though still appropriate) language, it will help the presentation to be more engaging, and being engaging was one of the key characteristics of effective presentations that we mentioned in Chapter 10. It will also help with using your voice (volume, pace, and pitch) as we mentioned in Section 11.1.4, 'Help the audience follow what you are saying'.

11.3.3 Think about your body language

Body language is an important and often neglected element of a presentation. It includes facial expressions and eye contact, as well as what you do with your hands, and where and how you situate yourself in the room. Body language can be both positive and negative—it can help a presentation be more engaging, or it can distract from or even counteract the presentation's content. If you have practised your presentation in front of a mirror, recorded yourself on your phone, or presented to your friends (as suggested in Chapter 10 'Preparing scientific presentations') you will be more aware of how your body language influences the way you present. Even though it can be counterproductive to become too self-conscious of your body language, here are some suggestions to make sure you use your body language positively.

Think about how you will use the available space

Where are you going to stand in the room (you might prefer to sit, but it's more normal to stand and it helps you project your voice better)? Where will you put your notes? If you are using a projection screen (with an overhead projector, slides, or Prezi), you need to be able to stand somewhere that doesn't obscure anyone's view of the screen, but still allows you to be close enough to move to the next element without getting in anyone's way or tripping over cables. Spend some time thinking about this before you start presenting because once you have started your presentation you want to be able to concentrate on what you are saying, rather than thinking about where the best place is for you to stand.

Think about what you will do with your hands

People use their arms and hands in everyday conversation to add emphasis or to help describe events. Presenters will therefore look rather awkward if they keep their hands in their pockets or rooted firmly at their sides. Equally, you don't want your hands to move in a way that distracts your audience's attention such as nervous, repetitive movements. Instead, use your hands to emphasize and enhance what you want to say verbally. As long as use of your hands is controlled and purposeful, then it has a key role to play in supporting your verbal communication. If you can try to relax a little and be conversational, then it is more likely that what you do with your hands will be more natural and more in sync with what you are saying.

Make eye contact

Often during an assessed presentation, the presenter only makes eye contact with the person doing the assessing, thereby excluding everybody else. If you were to have a conversation with someone you would make eye contact with them, and not doing so would probably be considered a bit rude. So when presenting, try to make eye contact with everyone. It is likely that you will only be presenting to a relatively small number of people, so making eye contact with everyone is a realistic aim. If, however, you are presenting to larger groups just try to make eye contact with all parts of the room. Making eye contact can be more difficult if you are using verbatim notes, so make sure you follow the guidelines on how to use verbatim or outline notes in Chapter 10, 'Preparing scientific presentations'.

Smile

If you are nervous it can be difficult to smile, but smiling can make a big difference to how people react to you. Clearly, you don't want to have an inane grin on your face for the entire presentation—that would be unnatural and look ridiculous—but making a point of trying to smile, at least before you start, will help your audience feel more at ease. It is likely that at least someone in the audience might smile back at you, which in turn helps you feel more at ease, too.

11.3.4 Show some enthusiasm!

Finally, show some enthusiasm! If you appear bored by and uninterested in what you are saying, your audience will probably react in the same manner. Conversely, if you can show genuine interest and enthusiasm, it is more likely that your audience will be willing to listen attentively and work at following what you are saying.

Try this—Practising your presentation

Get together with a small group of other students before you all have a presentation due. Each student should give their presentation, ideally in the room that you will be using. For each presentation, each member of the audience should write down one example of good practice and one area needed for improvement, and provide this written feedback to each presenter at the end of the session.

 Chapter summary

There are similarities between written and spoken presentations, but there are significant differences, too. It is often the differences that are the cause of concern or anxiety associated with delivering a presentation. This chapter has addressed these common concerns and has also focused on how to use a range of visual aids appropriately and the importance of mastering some key techniques. Following these guidelines will ensure that you deliver your scientific presentations much more effectively.

Creating academic posters

 Introduction

Poster presentations are a common academic format and are a regular feature at academic conferences. As an undergraduate, it is unlikely that you will be presenting a poster at a conference, but the poster format is sometimes used as a form of assessment for undergraduate courses. This chapter will explain the purpose of academic posters and suggest seven key steps for creating them effectively.

12.1 The context of poster presentations

If posters are used as a form of assessment at undergraduate level, then they have a different purpose from that of an academic conference. However, it is helpful to understand the context from which they come in order to know what they are supposed to achieve.

Academic conferences are important opportunities for researchers to present their work, allowing them to:

- publicize their findings;
- promote discussion;
- create opportunities for collaboration with other researchers.

Often there is not enough time for everyone at a conference to be able to present their research orally, so poster presentations are commonly used to allow lots of people to present their research at once. Typically, posters are size A1 (eight times bigger than A4) or A0 (16 times bigger than A4) and are attached to display boards arranged throughout an open plan room or conference hall. Time will be allocated in the conference schedule for delegates to view the posters, usually with opportunities to discuss the work with the presenter. A poster presentation is more limited than an oral presentation in terms of the amount of information that can be communicated and, therefore, posters should be designed to summarize the important elements and promote discussion.

Depending on how large the conference is and how well it is organized, delegates may have brief details of posters on display at the conference in advance, including the title of the poster and possibly an abstract. If these details are available, then delegates might scan the list of posters to identify those that particularly interest them. Alternatively (or additionally), delegates may simply browse the posters themselves and view those which catch their eye. Either way, the posters that are viewed the most will be those that stand out from the crowd in some manner.

12.2 Seven key steps to creating your poster

Now that we have explained the context of academic posters we need to identify how to create them effectively. There are seven key steps involved in the process; the first two relate to the content of the poster, the next four are about design issues, and the final one is about ensuring that it is free from errors. This may seem a little unbalanced but we have dealt a lot already in this book with writing for bioscience, and it is usually the design aspects, rather than the content aspects, where students need the most guidance. The seven steps, therefore, are:

1. Get the academic content right.
2. Cut the text down to between 300 and 500 words.
3. Format the type.
4. Prepare your images.
5. Design the layout.
6. Choose a colour scheme.
7. Check it very carefully.

12.2.1 Get the academic content right

First, you need to make sure that you get the academic content right. This is the foundation for everything else that follows. If you don't do this first you will waste a lot of time formatting and laying out content that you end up not using. Usually, for undergraduate assessment purposes, you will be given a specific brief for your poster (often based on another aspect of assessment, such as a practical report, essay topic, or research method). This brief will determine the parameters of the academic content. In addition to the academic content you may also be given guidance on the audience the poster is intended to be designed for. The intended audience will determine the level of the academic content that is appropriate.

Specialist

If the audience is from the same specialist field as the presenter, the presenter can assume that the people viewing the poster will have a high level of knowledge about the subject. It is, therefore, acceptable for the presenter to use technical language and terms on the poster.

Related

If the audience is from a field related to that of the presenter, the presenter can only assume that the people viewing the poster will be familiar with the generalities of the discipline. The presenter, therefore, needs to be careful about using technical language and terms.

General

If the audience is from the general public the presenter can only assume general knowledge. The people viewing the poster will probably have no familiarity with the discipline or subject area. The presenter, therefore, needs to avoid using technical language and terms, and should use only basic descriptions.

For the purposes of this chapter we will use the following brief for a poster.

> Design and produce an academic poster summarizing a current research technique in biological sciences. The intended audience is first-year undergraduate bioscience students. The poster needs to be size A1 and landscape orientation.

The current research technique we will use is *animal cloning*. We will follow this example through in the remaining six steps of the process.

12.2.2 Cut the text down to 300–500 words

The second step in creating your poster is to use an appropriate number of words. Clearly, from the title of this section we are assuming that initially you will have too many words, rather than too few. That's because a poster usually requires you to take a piece of work and summarize its content to the key details. At an academic conference the work will usually be some form of research; at an undergraduate level, as identified earlier, it will usually be based on another aspect of assessment, such as a practical report, essay topic, or research method. The recommended number of words for a poster is between 300 and 500; this may not seem like much (and it isn't!), but it is important to keep the number of words within these limits in order to create a poster that is both readable from a distance and conveys key information quickly. The number of words affects readability because the greater the number of words the smaller the font size needs to be in order to accommodate them on the page, and so the less readable it becomes. Also, if the poster contains too much text, the amount of information can be overwhelming to the reader, making it less likely that passing delegates will spend time viewing it. To get an idea of what 300 and 500 words looks like, see Figures 12.1 and 12.2, respectively.

When reducing the text down to between 300 and 500 words you need to remember that you are trying to make it easy for a person who is not familiar with the content to understand it quickly. It can be helpful to try to think of it as a bit like a trailer for a movie—you are not trying to tell people everything there is to know about the subject, rather you are just giving an overview to summarize the main points and encourage people to find out more.

For our poster on animal cloning we have summarized the current research techniques that are involved in the process and have tried to explain them appropriately for an audience of first-year undergraduate biological sciences students, as per the requirements of the brief. We have used a conventional scientific report pattern to provide a clear structure and have endeavoured to focus only on the essentials of animal cloning to create a succinct text. The words for our animal cloning poster are as follows:

Figure 12.1 A poster with 300 words

This poster contains 300 words

Lorem ipsum dolor sit amet, consectetuer adipiscing elit. Aenean id libero. Nam a lorem ac metus tempor iaculis. Cras eros velit, ullamcorper sit amet, semper fringilla, cursus ut, enim. Mauris eget elit. Nulla orci.

Morbi id neque. Ut lorem. Nulla euismod. Lorem ipsum dolor sit amet, consectetuer adipiscing elit. Fusce dictum, velit id commodo rhoncus, est orci elementum risus, a euismod eros sem ut magna.

Nulla accumsan orci quis est. Nunc nec massa vitae risus porttitor iaculis. Sed cursus felis congue nisi. Sed consequat. Nulla eros nunc, vulputate sed, ultrices nec, varius eu, arcu. Fusce a eros. Nulla hendrerit ornare ligula.

[IMAGE]

[IMAGE]

Vivamus pellentesque eros. Ut lobortis. Nullam bibendum. Nam ullamcorper ultrices mi. Mauris et pede at turpis mattis feugiat. Aliquam sit amet odio eu est mollis accumsan.

Fusce mattis pede sed orci. In imperdiet ante ut lorem. Aliquam tincidunt erat non sem. Pellentesque auctor nisl nec metus. Pellentesque habitant morbi tristique senectus et netus et malesuada fames ac turpis egestas. Sed commodo vulputate elit. Curabitur nonummy tincidunt felis. Pellentesque rhoncus lorem non leo.

Etiam a arcu. Quisque iaculis mi sit amet orci. Mauris malesuada massa aliquam nunc egestas dignissim. Nullam in nibh. Aliquam aliquam, nisl vel suscipit dignissim, urna urna dictum libero, sed cursus sapien nibh non mi.

Cras lorem lectus, volutpat ac, facilisis vel, consectetuer a, enim. Donec molestie laoreet erat. In tincidunt. Vivamus at massa. Curabitur eget augue sit amet est porta elementum. Lorem ipsum dolor sit amet, consectetuer adipiscing elit. Phasellus dui. Ut in pede et massa ullamcorper imperdiet. Donec ut nibh ut diam mollis faucibus. Fusce sit amet est. Duis tortor. Mauris sit amet purus.

Fusce pulvinar leo quis orci. Maecenas eget tortor vitae magna adipiscing eleifend. Suspendisse potenti. Curabitur pretium sagittis arcu. Nam libero. Sed nec lacus id leo dictum scelerisque. Cras interdum nisi id.

Figure 12.2 A poster with 500 words

This poster contains 500 words

Lorem ipsum dolor sit amet, consectetuer adipiscing elit. Mauris eu urna vitae risus feugiat lacinia. Duis purus turpis, hendrerit ut, vestibulum eget, pulvinar nec, purus. Donec vitae nibh nec eros ultrices eleifend. Nullam sed elit ac quam iaculis pulvinar. Phasellus sit amet leo.

Sed metus quam, hendrerit vel, sollicitudin vitae, eleifend vitae, lorem. Sed nulla pede, sollicitudin id, lobortis ac, venenatis eu, mauris. Aenean molestie, leo quis elementum laoreet, urna dui iaculis enim, sit amet viverra nisl lorem elementum erat. Etiam pharetra. Donec odio massa, rhoncus eget, egestas quis, mattis lobortis, massa. Cras nonummy porta nibh. Vivamus ornare imperdiet nisl. Suspendisse potenti. Nullam ullamcorper. Fusce eget orci. Sed tortor. Nulla sit amet libero iaculis metus porta pulvinar. Maecenas ultricies, mi ut nonummy gravida, orci nunc porttitor orci, non pharetra urna massa eget turpis.

[IMAGE]

Nam ligula sapien, volutpat eu, elementum at, imperdiet a, sapien. Fusce eu sapien. Nullam venenatis tortor a ligula. Integer sed quam nec odio hendrerit pretium. Suspendisse nulla arcu, venenatis sed, viverra quis, molestie euismod, enim. Morbi odio. Pellentesque vestibulum risus nec eros. Suspendisse lacinia ornare nunc. Etiam aliquam enim sed nisl. Sed lacus lorem, placerat eget, accumsan at, pretium sit amet, tellus. Fusce ut elit non felis sagittis luctus. Class aptent taciti sociosqu ad litora torquent per conubia nostra, per inceptos hymenaeos. Integer molestie odio posuere pede. Aenean lacus. Suspendisse tellus.

Aliquam dapibus. Cras dolor neque, posuere nec, nonummy eget, vulputate eu, augue. Suspendisse potenti. Nullam non mi faucibus est consectetuer imperdiet. Quisque dignissim volutpat tortor. Nulla accumsan. Cras accumsan, libero vel cursus placerat, felis orci pharetra massa, a volutpat felis purus id metus. Cum sociis natoque penatibus et magnis dis parturient montes, nascetur ridiculus mus. Duis semper mauris ac tortor. Praesent eu ipsum. Sed sed arcu. Quisque vestibulum pellentesque nisl. Curabitur egestas posuere lectus. Pellentesque enim. Nunc vestibulum, est ac mattis commodo, mass alectus tincidunt est, vitae bibendum velit urna in nisi.

Praesent odio mi, consequat aliquet, rhoncus eget, congue vitae, velit. Suspendisse volutpat, urna id malesuada rhoncus, nisl lorem cursus leo, in venenatis dui nibh vel elit. Vivamus quis lorem iaculis magna commodo nonummy. Nulla enim justo, viverra nec, nonummy vel, ullamcorper sit amet, odio. Aliquam elementum nonummy leo.

Nam bibendum risus. Class aptent taciti sociosqu ad litora torquent per conubia nostra, per inceptos hymenaeos. Nunc in lacus. Nam et dolor. Nulla ullamcorper. Sed nec diam ac tellus tempus viverra. Vestibulum posuere pede eleifend est. Vivamus leo nibh, aliquet suscipit, vehicula non, condimentum nec, nulla. Cum sociis natoque penatibus et magnis dis parturient montes, nascetur ridiculus mus. Ut mauris. Pellentesque habitant morbi tristique senectus et netus et malesuada fames ac turpis egestas. Pellentesque auctor quam pulvinar risus. Pellentesque dignissim pretium eros. Aliquam id ante.

Quisque bibendum mauris vitae tortor. Nam pulvinar lorem ut nulla. Donec felis. Suspendisse gravida, lectus id imperdiet sagittis, purus ante blendit arcu, non ultrices nisi metus ut nulla. Proin vel velit et turpis feugiat lobortis. Etiam tristique, dui et dignissim dapibus, nisl purus consectetuer sapien, eu viverra metus dolor at odio.

Introduction

Animal cloning is a contentious topic. The history of cloning dates back to the nineteenth century (Driesch, 1892); however, animal cloning came to particular public prominence in 1997 when Wilmut *et al.* demonstrated that the cloning of mammals from adult cells, rather than an embryo, was possible. The authors of the study derived nuclei of cultured mammary epithelial cells from an adult ewe, resulting in the birth of 'Dolly'—the first cloned sheep.

Methods

Cloning from adult cells involves taking the nucleus of a cell from the animal intended to be cloned and transferring it into an unfertilized ovum, which has had its metaphase chromosomes removed. If this step is successful, the transferred nucleus is 'reprogrammed' to initiate the development of a new embryo. This embryo is genetically identical to the animal from which the cell was obtained. The embryo is then transferred into a surrogate mother for gestation to term and birth.

However, there are many factors affecting the success of cloning, including: the type of nuclear donor cell used; the treatment of donor cells prior to nuclear transfer; the source of recipient ova; the techniques employed; and the competency of the laboratory group working on it.

Cloning in cattle

There are more laboratories worldwide cloning cattle than all other species combined. This is partly due to the knowledge and expertise of assisted reproductive techniques in cattle developed over the past 20 years. While the success and repeatability of cloning from embryonic cells are now well established in cattle, cloning from adult cells still remains more problematic.

Cloning in pigs

The cloning of pigs has illustrated well the fact that assisted reproductive technologies and techniques for nuclear transfer do not directly apply from one species to another. In a study by Onishi *et al.* (2000), the transfer of 110 cloned embryos to four surrogate mothers produced only one piglet.

Results

Table 1 shows the success rates of animal cloning from mammalian adult cells. The figures are taken from a review article by Solter (2000).

It is evident from the above figures that, even given the significant progress made in the field of animal cloning from adult cells in recent years, success rates are still very low.

Conclusion

Techniques used to clone animals are still developing, and cloning to reproduce certain genotypes can represent a significant challenge. With a variety of cloned animals being

(Continued)

Table 1 Cloning efficiency

Species	Number of oocytes manipulated	Success rate (%)
Cow	99	2
Pig	401	1.2
Mouse	1345	0.7
Sheep	277	0.4

produced, the demand for cloning specific animals will increase significantly in the coming years. Differences in the ability to clone a particular species are a result of the time and resources that have been invested into research.

References

Driesch, H. 1892. Entwicklungsmechanische Studien. I. Der Wert der beiden ersten Furchungszellen in der Echinodermentwicklung. *Zeitschrift für Wissenschaftliche Zoologie*, 53, 160–178.

Onishi, A., Iwamoto, M., Akita, T., Mikawa, S., Takeda, K., Awata, T., Hanada, H., Perry, A. 2000. Pig cloning by microinjection of fetal fibroblast nuclei. *Science* 289(5482), 1188–1190.

Solter, D. 2000. Mammalian cloning: advances and limitations. *Nature Reviews Genetics* 1, 199–207.

Wilmut, I., Schnieke, A.E., McWhir, J., Kind, A.J., Campbell, K.H. 1997. Viable offspring derived from fetal and adult mammalian cells. *Nature*, 385(6619), 810–813.

The text contains 462 words (not including the references). The structure follows a conventional scientific report pattern: Introduction, Methods, Results, Conclusion, and References.

12.2.3 Format the type

Thirdly, you need to format the type. When formatting type for a poster, you need to remember two important principles: to use consistent styles, and to group sections of text appropriately.

Use consistent styles

Using consistent styles helps achieve what graphic designers call the principle of *resemblance*. It means that text which performs the same function should always be formatted in exactly the same way. For instance, if you decide that the subheadings should be Arial, 10 point, bold, left-hand justified with an after paragraph line spacing of 0.2 lines, then all your subheadings should be formatted in exactly this way. Similarly, you might decide that the captions for figures should be Arial, 6 point, regular, right-hand justified, with a before and after paragraph line spacing of 0.2 lines, in which case all your captions should be formatted in exactly this way (see Box 12.1).

There are two reasons for using consistent styles: first, it makes the poster look neater; secondly, it makes the structure of the poster easier to understand, because you know which text is performing what function. An example of using consistent styles is shown in Figure 12.3.

Figure 12.3 Using a limited number of consistent styles

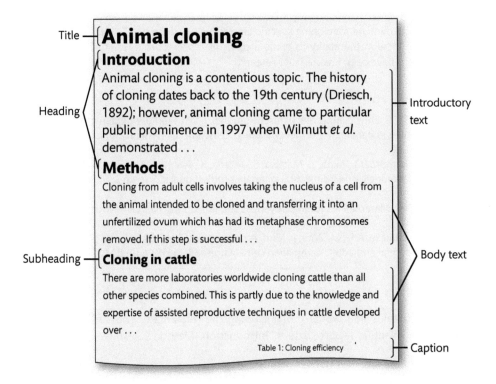

BOX 12.1 How do I know what font size to use?

Posters are often created at A4 size and then enlarged to the appropriate size for a poster, usually either A1 or A0. This can make it difficult to know what font size is appropriate to use because you are not creating your poster at the size of the finished product. The magnification factor from A4 to A1 is approximately 300% (it is actually 283%, for reasons that we won't go into here) and from A4 to A0 is 400%. This relationship is illustrated for you in Figure 12.4.

Therefore, a font size of 10 point at A4 is equivalent to a font size of 40 point at A0. Approximate appropriate font sizes to use for A1 and A0 posters are shown in Table 12.1.

(Continued)

Table 12.1 Suggested font sizes to allow appropriate magnification

Style	A4 to A1	A4 to A0
Title	18 point Arial bold	16 point Arial bold
Heading	13 point Arial bold	12 point Arial bold
Subheading	11 point Arial bold	10 point Arial bold
Introductory text	11 point Arial bold	10 point Arial regular
Body text	10 point Arial bold	8 point Arial regular
Captions	8 point Arial bold	6 point Arial regular

Figure 12.4 A4 paper to A1 or A0

Figure 12.5 Inappropriate proximity of text

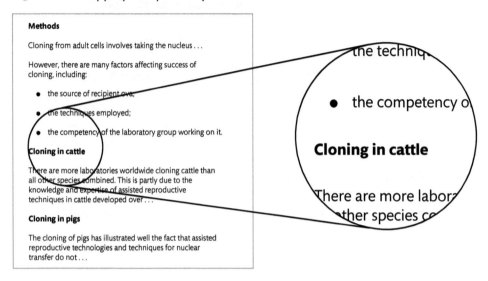

Group sections of text appropriately

Grouping sections of text appropriately helps achieve what graphic designers call the principle of *proximity*. It means that text which belongs together is grouped together; this is usually achieved using paragraph spacing. Figure 12.5 illustrates inappropriate spacing of text; there is a whole line space between each section and in the bulleted list. The result is that the headings and bullets float around and it is not immediately clear which piece of text belongs to which section.

Figure 12.6 illustrates more appropriate spacing of text; paragraph spacing has been used to make it clear which pieces of text the headings and bullets belong to.

Figure 12.6 Appropriate proximity of text

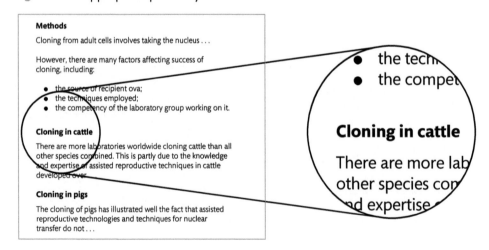

Figure 12.7 Line spacing and body text

> **Line spacing = 0.9**
> It is evident from the above figures that, even given the significant progress made in the field of animal cloning from adult cells in recent years, success rates are still very low.
>
> **Line spacing = 1.0**
> It is evident from the above figures that, even given the significant progress made in the field of animal cloning from adult cells in recent years, success rates are still very low.
>
> **Line spacing = 1.1**
> It is evident from the above figures that, even given the significant progress made in the field of animal cloning from adult cells in recent years, success rates are still very low.

Line spacing, alignment, and text formatting

In addition to thinking about the principles of resemblance and proximity when formatting the type, it is also important to consider line spacing, alignment, and text formatting. Figure 12.7 illustrates the effect of line spacing on body text; body text (which is usually the smallest text on the page other than captions) is much easier to read if the line spacing is slightly increased.

Titles that run to more than one line, however, often benefit from the reverse: compressed line spacing, as shown in Figure 12.8. However, where possible, it is better if you can just make your titles more succinct so that they can fit on one line.

It is recommended that you left-hand justify all body text (as shown in Figure 12.9) because fully justifying the body text sometimes results in awkward gaps on some lines, making the text more difficult to read (as shown in Figure 12.10).

Figure 12.8 Line spacing and titles

Large text (e.g. titles) probably needs decreased line spacing . . .

Line spacing = 0.9

. . . because it looks a bit strange with increased line spacing.

Line spacing = 1.1

Figure 12.9 Left-hand-justified body text

> Animal cloning is a contentious topic. The history of cloning dates back to the 19th century (Driesch, 1892); however, animal cloning came to particular public prominence in 1997 when Wilmut *et al.* demonstrated that the cloning of mammals from adult cells, rather than an embryo, was possible. The authors of the study derived nuclei of cultured mammary epithelial cells from an adult ewe, resulting in the birth of 'Dolly'—the first cloned sheep.

Figure 12.10 Fully justified body text

> Animal cloning is a contentious topic. The history of cloning dates back to the 19th century (Driesch, 1892); however, animal cloning came to particular public prominence in 1997 when Wilmut *et al.* demonstrated that the cloning of mammals from adult cells, rather than an embryo, was possible. The authors of the study derived nuclei of cultured mammary epithelial cells from an adult ewe, resulting in the birth of 'Dolly'—the first cloned sheep.

Finally, on formatting the type:

- use the same font throughout your poster (unless using a complementary font for headings);
- set the headings in bold;
- use *italics*, underlining, and CAPITALS sparingly (only when convention dictates);
- break up any large areas of text with subheadings.

12.2.4 Prepare your images

The fourth step is to prepare your images. Images can be photographs, diagrams, graphs, or tables. The appropriate use of images can benefit a poster considerably, not just in terms of its visual impact, but also in terms of how readily understandable the content of the poster is. We will deal with each of these image types in turn.

Photographs

There are three key points you need to remember about any photographs you use on a poster.

1. They need to be relevant to the content of your poster, not just put in to brighten it up.
2. If they are not your own photographs they need to be appropriately referenced (in the same way as you would reference text that wasn't your own).
3. They need to be of an appropriate resolution.

In order for photographs in particular, and images generally, not to degenerate, they need to be of sufficient quality to start with—that is, they need to be of a sufficiently high resolution. The resolution of digital images is measured in pixels per square inch. Photographs copied from the internet usually have a resolution of 72 pixels per inch and so will look 'grainy' or 'pixelated' when enlarged, as shown in Figure 12.11.

For a printed poster, choose images that have a resolution of at least 300 pixels per inch at the size at which they will appear on the poster. In all likelihood you will want to enlarge the image when you reproduce it for the poster: and as soon as you enlarge it, the resolution will decrease. For example, if you have an image that is 2 cm square and has a resolution of 300 dpi, when you enlarge it to become 4 cm square, its resolution will drop to 150 dpi (which is

Figure 12.11 The effect of enlargement on low-resolution images

too low a resolution for printing purposes). Instead, start with an image that has a resolution that's well above 300 dpi. Then, when you enlarge it, the resolution will drop—but hopefully not below 300 dpi.

Diagrams

A diagram can be a very useful way of communicating information in a quick and succinct manner. Therefore, they are of particular use in posters, where communicating information in a readily understandable and concise fashion is very important. A well-chosen diagram can save a lot of words or make text descriptions much easier to understand. In our example poster, Figure 12.12 supplements the information in the first paragraph of the methods section to make it more readily understandable.

Graphs

When making graphs for posters, you need to think carefully about how you format them, as viewing them from a distance means that default formatting is usually not appropriate. The graph in Figure 12.13 is difficult to view from a distance because of the relatively small font, the shaded plot area and the excessive use of horizontal lines. All these formatting features are default features in Microsoft Excel. As a general rule, when it comes to the formatting of a graph, don't accept the default formatting provided by your software.

Figure 12.12   A diagram saving a lot of words

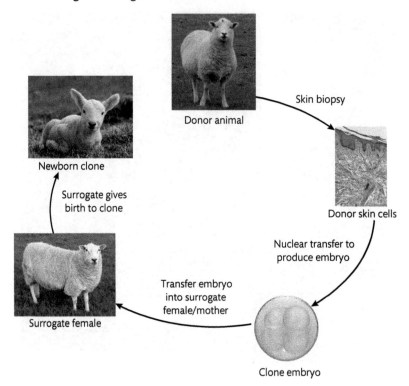

Lamb: Donald Macleod. licensed under the Creative Commons Attribution 2.0 Generic license. *Donor sheep*: this file is licensed under the Creative Commons Attribution-Share Alike 3.0 Unported license. *Embryo*: © Istock/Gary Caviness. *Skin*: © Karl Dolenc/istock.

Figure 12.13   An inappropriately formatted graph

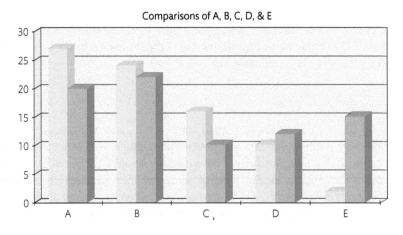

Figure 12.14 A more appropriately formatted graph

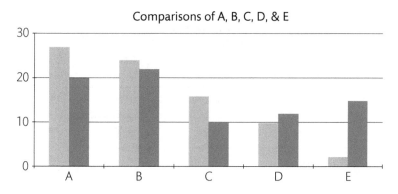

Figure 12.14 shows a more appropriately formatted graph. The font size has been increased, the shading on the plot area has been removed, and the use of horizontal lines has been minimized. Additionally, the colours of the bars have been altered to provide greater contrast; making the two sets of figures easier to distinguish between. Again, as with photographs and diagrams, if the data are not your own you need to reference them appropriately.

Tables

Finally, in this section on preparing images, let's consider tables. The principles that applied to graphs apply equally to tables—format them in such a way so as to make them clear, simple, and easy to see. Again, default formatting is usually inappropriate, as it tends to make the text too small and the grid lines can make patterns in the data difficult to discern. As a general rule, don't format tables that look like spreadsheets, as shown in Figure 12.15, as they are more difficult to interpret, even when you are dealing with a small data set.

Figure 12.16 shows a table that has been formatted more appropriately. The following changes have been made:

- the font size has been increased;
- the header row is made clear by setting the text in bold and lightly shading the row;

Figure 12.15 Default formatted table

Species	Number of oocytes manipulated	Success rate (%)
Sheep	277	0.4
Cow	99	2
Mouse	1345	0.7
Pig	401	1.2

Figure 12.16 A more thoughtfully formatted table

Species	Number of oocytes manipulated	Success rate (%)
Cow	99	2.0
Pig	401	1.2
Mouse	1345	0.7
Sheep	277	0.4

- the internal vertical lines have been deleted to allow patterns between the rows to be more easily seen;
- the numbers have all been right-hand justified so that the decimal place is vertically aligned;
- the percentages have all been quoted to the same number of decimal places;
- the rows have been ordered by the magnitude of the aspect of the data that attention needs to be drawn to: the percentage success rate.

12.2.5 Design the layout

The fifth step is to design the layout. The layout needs to be designed so that a person viewing your poster can quickly understand the sequence of the information. This is most easily achieved using appropriate and conventionally ordered headings, as shown in Figure 12.17.

In addition to using a conventional heading sequence it is also helpful to use a design grid, as shown in Figure 12.18. A design grid is simply a series of horizontal and vertical lines on the page that are used for lining sections up; the lines are then removed when the poster is printed. Most drawing software has a design grid feature, or alternatively you can draw the lines manually with a line tool to create the grid and remove them before printing.

In the previous examples, the methods section is unusually long. However, remember that this is a poster summarizing a current research technique. More commonly, the methods section would be much shorter, leaving space for fuller results, discussion, and conclusion sections. In spite of the unusually long methods section, however, the layouts are fairly conventional, following a symmetrical three-column format. There are many other alternatives though, and you will need to choose the most appropriate one to fit your content. Other examples for both landscape and portrait layouts are shown in Figures 12.19 and 12.20.

12.2.6 Choose a colour scheme

The penultimate step is to choose a colour scheme. Try to use only two or three different colours, plus black, which is always best for the smallest text. There are two ways of choosing your colours: you can use colours from your images or a colour wheel.

Figure 12.17 Use of appropriate headings to show sequence

Animal Cloning
Stuart Johnson and Jon Scott

Introduction

Methods

Methods
(continued)

Results

Species	Number of oocytes manipulated	Success rate (%)
Cow	99	2.0
Pig	401	1.2
Mouse	1345	0.7
Sheep	277	0.4

Table 1: Cloning efficiency

Conclusion

References

skin biopsy
Donor animal
Newborn clone
surrogate gives birth to clone
Donor skin cells
Surrogate female
Clone embryo
Figure 1: Cloning by nuclear transfer

Cow: Keith Weller/USDA.

Colours from images

One way of selecting colours is to use one that is represented in your images. So for the poster we have used as an example we might choose a pink because of the pig photograph, or a brown because of the cow photograph.

Colour wheel

The alternative to choosing colours from images (either because you don't have any appropriately coloured images or you are not confident in choosing from them) is to choose colours from a colour wheel. A colour wheel shows the primary, secondary, and tertiary colours, and can be used to choose appropriate mixes of colour.

You can choose either:

- analogous colours (colours next to each other on the colour wheel);
- complementary colours (colours opposite each other on the colour wheel);
- or shades (different shades of the same colour).

Figure 12.18   Use of a design grid to show sequence

Animal Cloning

Stuart Johnson and Jon Scott

Introduction

Animal cloning is a contentious topic. The history of cloning dates back to the 19th century (Driesch, 1892); however, animal cloning came to particular public prominence in 1997 when Wilmut et al. demonstrated that the cloning of mammals from adult cells, rather than an embryo, was possible. The authors of the study derived nuclei of cultured mammary epithelial cells from an adult ewe, resulting in the birth of 'Dolly'—the first cloned sheep.

Methods

Cloning from adult cells involves taking the nucleus of a cell from the animal intended to be cloned and transferring it into an unfertilized ovum which has had its metaphase chromosomes removed. If this step is successful, the transferred nucleus is 'reprogrammed' to initiate the development of a new embryo. This embryo is genetically identical to the animal from which the cell was obtained. The embryo is then transferred into a surrogate mother for gestation to term and birth (see Figure 1).

Figure 1: Cloning by nuclear transfer

However, there are many factors affecting success of cloning, including:

- the type of nuclear donor cell used;
- the treatment of donor cells prior to nuclear transfer;
- the source of recipient ova;
- the techniques employed; and
- the competency of the laboratory group working on it.

Cloning in cattle

There are more laboratories worldwide cloning cattle than all other species combined. This is partly due to the knowledge and expertise of assisted reproductive techniques in cattle developed over the last 20 years. Whilst the success and repeatability for cloning from embryonic cells is now well established in cattle, cloning from adult cells still remains more problematic.

Cloning in pigs

The cloning of pigs has illustrated well the fact that assisted reproductive technologies and techniques for nuclear transfer do not directly apply from one species to another. In a study by Onishi et al. (2000) the transfer of 110 cloned embryos to four surrogate mothers produced only one piglet.

Results

Table 1 shows the success rates of animal cloning from mammalian adult cells. The figures are taken from a review article by Solter (2000).

Species	Number of oocytes manipulated	Success rate (%)
Cow	99	2.0
Pig	401	1.2
Mouse	1345	0.7
Sheep	277	0.4

Table 1: Cloning efficiency

It is evident from the above figures that, even given the significant progress made in the field of animal cloning from adult cells in recent years, success rates are still very low.

Conclusion

Techniques used to clone animals are still developing, and cloning to reproduce certain genotypes can represent a significant challenge. With a variety of cloned animals being produced, the demand for cloning specific animals will increase significantly in the coming years. Differences in the ability to clone a particular species are a result of the time and resources which have been invested into research.

References

- Driesch H 1892. Entwicklungsmechanische Studien. I. Der Wert der beiden ersten Furchungszellen in der Echinodermentwicklung. *Zeitschrift fur wissenschaftliche Zoologie* 160–178.
- Onishi A, Iwamoto M, Akita T, Mikawa S, Takeda K, Awata T, Hanada H, Perry A 2000. Pig cloning by microinjection of fetal fibroblast nuclei. *Science* **289**(5482), pp. 1188–1190.
- Solter D 2000. Mammalian cloning: advances and limitations. *Nature Reviews Genetics* 1, pp. 199–207.
- Wilmut I, Schnieke AE, McWhir J, Kind AJ, Campbell KH 1997. Viable offspring derived from fetal and adult mammalian cells. *Nature* **385**(6619), pp. 810–813.

Figure 12.19   Alternative landscape layout

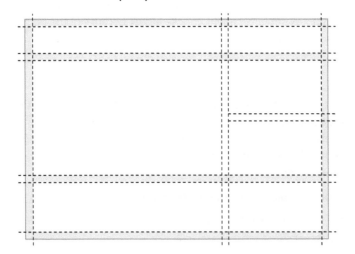

Figure 12.20 Alternative portrait layouts

12.2.7 **Check it very carefully**

Finally, you need to check your poster very carefully (see Box 12.2). This is particularly important if you are going to get an enlarged, laminated copy printed: you don't want to spend money printing it out only to find the errors afterwards. A good way to check your poster is to print it out A4 (preferably in colour), stick it on the wall, and take a step back to look at it.

BOX 12.2 Do the small details really matter?

Posters can be very fiddly and therefore time-consuming to create. However, remember the academic context of poster presentations that we summarized in Section 12.1, 'The context of poster presentations': in a busy conference setting with lots of posters to view, people make very quick (often unfair) judgements about whether or not they will bother viewing a poster. We said that a poster needs to stand out from the crowd in some way if it is going to be viewed. Positively, this will include a snappy title, good visual impact, and information that is readily understandable. Negatively, small errors in the design can put people off and lead them to think that if the design isn't accurate the data might not be either. Also, if you are creating a poster at A4 size, then enlarging it to A1 or A0, any small errors at A4 will be enlarged 300% or 400%, respectively. Therefore, attention to detail is important. See Figure 12.21 for an example of how small inaccuracies with the layout at the A4 stage lead to much larger inaccuracies at A0.

(Continued)

Figure 12.21 How small errors are noticeable when the design is enlarged to poster size

This will simulate more closely how the real thing will be viewed and help you find any errors more easily. Alternatively, if you have access to a data projector, you can project the image onto a screen to see the effect of the image being blown up to full size (although the image resolution is much less for digital images than printed images, so don't use a projected image as a guide for image quality).

 Chapter summary

Designing academic posters can be a difficult and time-consuming task. However, by understanding the purpose of posters and the key steps involved in creating them, designing and producing quality academic posters becomes an achievable goal. Designing posters is a particularly useful skill if you continue studies beyond undergraduate level, as they are a common form of presenting research at academic conferences.

Getting the most out of revision

 Introduction

The results you gain for your degree will depend on how you perform in the various types of assessment that you are required to undertake. We have already addressed essays, practical and project reports, presentations, and posters; the one we haven't dealt with yet directly (and the one that for most degrees accounts for the majority of the marks) is examinations. In order to do well in exams you have to do well at revision (two words that usually conjure up fairly negative emotions in people). This chapter will show you how you can revise well; Chapter 14, 'Getting the most out of exams', will deal with the exam skills.

If exams account for the majority of the marks that make up your degree result, it is understandable that people feel pressured or stressed by them. One important way of significantly increasing your chances of doing well, and of alleviating some of the pressure and stress of exams is to get the revision part right. If you can go into an exam confident that you have prepared yourself as well as you could have done (or, at least, reasonably well), then a significant amount of the pressure and stress can be eased, making it more likely that you will perform well. However, getting the revision part right takes a significant amount of organizing; it is no good leaving it to the last minute. Revising for exams is a bit like training for a marathon: if you want to do well you can't cram all your training into the last few days—you've got to make a sensible training plan for the weeks and months beforehand, and then do your best to stick to it.

First though, let's consider what revision actually is. In very general terms, revision is to do with going over a subject again, usually in preparation for some kind of exam. However, exactly how you go over a subject again is critical to whether or not the revision is effective. You will know if your revision is effective because revision that is effective will make the material you are revising easier to:

- understand;
- remember;
- apply.

This chapter will give you some tips and techniques to help you make this happen.

13.1 Get yourself organized!

We have already highlighted the fact that getting revision right takes a significant amount of organizing. However, getting yourself organized is easy to say, but is more difficult actually to do, especially when you're beginning to feel the pressure of exams looming and you just want to get on with revising. It is important, however, to see the organizing or planning aspects of revision as part of the process, rather than as something to get out of the way so you can get on with the 'real work'. This is similar to how we identified planning writing as an important part of writing essays in Chapter 7, 'Writing essays'. Time spent *organizing* revision is time well spent (as long as you don't use it as an excuse to procrastinate!)

There are several questions you need to ask yourself in order to get organized.

- What will I be examined on?
- Do I have all the necessary material?
- How much time do I have before the exams?
- How am I going to allocate my time?

These questions will form the structure of this first section.

13.1.1 Find out what you will be examined on

First, then, you need to find out as much as possible about what you will be examined on. This includes both the format and scope of the exam (or exams). In terms of format this would include finding out answers to the following questions.

- How long will the exam last?
- How many questions will there be?
- Will you have any choice about which questions you answer?
- What type of questions will you be asked (e.g. multiple-choice, short-answer, or essay)?

Knowing the answers to these basic questions will help you to know what to expect. The better informed you are, the better you can prepare. It also means that you are not surprised when you open the exam paper and so will be more able to concentrate on answering the questions, rather than spending time trying to understand issues of format.

In addition to the format of the exam you also need to consider its scope. What are you expected to know in terms of both breadth (the range of information) and depth (the amount of detail)? Imagine you were revising for an exam on comparative animal physiology: if you had revised enough in terms of breadth, but had revised insufficient depth, your knowledge of comparative animal physiology would be too *shallow*. For example, you may have revised the comparative physiology of the respiratory, cardiovascular, and excretory systems, ion balance, and vision, but failed to learn about them in sufficient detail. Alternatively, if you had revised enough in terms of depth, but had revised insufficient breadth, your knowledge of comparative animal physiology would be too *narrow*. For example, you may have revised vision in intricate detail, but if you failed to learn about the other aspects as well (respiratory, cardiovascular, and excretory systems, and ion balance), then you would be very limited in the number of questions you could answer.

This may all seem fairly obvious, but finding out what you will be examined on is an important and often overlooked area of revision. Sometimes it is overlooked due to carelessness, other times due to a deliberate, but irrational avoidance mentality. Whatever the reason, it is not difficult to fix. There are a number of ways you can find out information about exam format and scope including:

- module handbooks;
- past papers;
- course tutors.

Each of these will give you important insights into what you can expect.

13.1.2 Check you have all the necessary material

The second question to ask to get yourself organized is 'Do I have all the necessary material?' When you have found out about what you will be examined on (the breadth and depth), you then need to look at the material you have, and decide whether or not you have enough. It is important that you think about more than just your lecture notes at this point. Other important sources of information include:

- notes from tutorials;
- essays you have written for coursework;
- lab reports that you have submitted;
- notes from presentations that you have been required to give;
- notes from additional reading for all of the above;
- feedback from tutors on all of the above.

As you can see, the amount of information you will have accumulated over a term or semester will be considerable. At this stage, you are not supplementing your notes, simply auditing them to check where there are any gaps. We said earlier that module handbooks were an important source of information on exam format and scope, and as such they help you to identify gaps in your notes and so help you determine whether or not you have enough material. Simply make a list of what information you have on a summary sheet and compare it to what you are expected to know, as described in the learning outcomes for the particular module you are revising for.

When you identify gaps in your material (we say 'when', rather than 'if' because it is unlikely that you will have everything), make a note of what the gaps are so you can come back to them later. It is also worth identifying whether the gap is a breadth issue (an insufficient range of information) or a depth issue (insufficient detail). The other factor to consider is whether certain gaps actually matter—if the format of the exam allows you some choice as to which questions you answer (e.g. some essay-based exams) you may not need to revise everything. It is usually better to have revised most things in sufficient depth, rather than everything at just a surface level. It may also be that there were some sections of the module or course that you just could not get to grips with and still don't understand. There are risks associated with this strategy of selective revision, so you need to be careful. For essay-based exams where you have a choice of questions, it is still very important that you revise more topics than you actually have to answer questions on, so that you do have a choice when you come to read the paper. A reasonable rule of thumb is always to learn at least two more topics than you have to answer questions on, so if, for example, your typical essay paper requires you to answer three questions, then it is best to know five topics really well.

13.1.3 Note how much time you have before the exams

The third aspect of getting yourself organized is to be clear about how much time you have to revise. Once you know the dates of your exams you will know how much time is available to you. The best way to impress on yourself how much time you have available is to represent the time visually by using a diary, calendar, or simply a sheet of paper. In addition to the exam dates themselves, you need to include other fixed points, such as any remaining coursework deadlines, the final lecture, and any non-course commitments you have, such as holidays. All of this needs to be represented on a planner, such as the one shown in Figure 13.1.

Although your exams are likely to be scheduled within a defined period, it is likely that there will be some gaps between them, so you can stagger the start of the revision for each exam accordingly. Knowing when your exams are will help you to appreciate how much time you have before the exams. The next step is to decide how you are going to allocate the time you have.

13.1.4 Decide how you are going to allocate your time

Deciding how to allocate your time is an important element of getting yourself organized. The following guidelines will help.

Decide what you are going to revise and how much

In order to decide how you are going to allocate time for revision you need to decide what you are going to revise and how much you are going to revise it. You will already have some idea of this, having found out what you will be examined on (Section 13.1.1, 'Find out what you will be examined on') and whether you have all the necessary material (Section 13.1.2, 'Check you have all the necessary material'), but now you need to make some definite decisions about how you will fit the revision of your material to the time you have available. There are some important considerations to bear in mind at this point:

- if you are going to know your material in sufficient depth you will need to revise subjects more than once;
- however, that doesn't mean you have to revise absolutely everything (remember, it is usually better to have revised most things in sufficient depth, rather than everything at just a surface level);
- you might also decide that you need to revise more for some exams than for others—you don't necessarily have to allocate equal amounts of revision time to each exam.

Figure 13.1 Planning your time

April						
		1	2	3	4	5
6	7	8	9	10	11	12
13	14	15	16 Final lecture	17	18	19
20	21	22 Essay due in	23	24	25 Away for weekend	26
27	28	29	30			

May						
				1	2	3
4	5	6	7	8	9	10
11	12	13	14 Exam 1	15	16 Exam 2	17
18	19	20 Exam 3	21 Exam 4	22	23	24
25	26	27	28	29	30	31

A useful process is to make an overview of each subject, listing the topics in that subject, and the headings under each topic (you will already have most of this information if you have identified whether you have the necessary material). This creates an index of what needs to be revised so that you can divide the revision into easy-to-manage sections. Once you have this overview, you are ready to put together your revision timetable.

Make a revision timetable

The kind of plan represented in Figure 13.1 gives you a useful overview, but it doesn't give you enough detail about how you need to allocate your time. You need to zoom in on this overview to create monthly, weekly, and even daily plans of what you need to do.

Imagine your four exams were as follows:

1. Comparative animal physiology: 14 May
2. Human genetics: 16 May
3. Biochemistry of enzymes: 20 May
4. Animal and plant diversity: 21 May

According to your plan (Figure 13.1) this gives you four weeks between your final lecture (16 April) and your first exam (14 May). However, there is also your final essay to submit on 22 April, and—assuming you will need the time between the final lecture and the submission date to work on the essay—that leaves just over three weeks to revise before your first exam. The remaining three exams then fall within the following seven days. These weeks need to be carefully planned if you are to cover the necessary material in the time available; this is where the weekly and daily plans come in (Figure 13.2).

Try this—Planning your revision

Two months before your next set of examinations, prepare a detailed revision timetable. Make it realistic by taking into account other commitments that you might have, such as work, social events, and family time.

Be realistic and build in some slack

It is important to be realistic in your planning. Estimating clearly how much time you have should help you not to panic and so end up trying to cram too much revision into a short space of time. If you are clear about how much time you have and what you need to do, you can then allocate your time realistically. A realistically planned revision timetable will have some slack built in so that when your plans need to change (for instance, due to a certain

Figure 13.2 Monthly, weekly, and daily revision plans

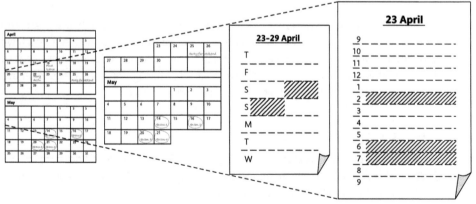

Figure 13.3 Daily plan with slack built in

topic taking longer than expected) there is space for this to happen. Of course, if the slack time is not needed you can use it for other things, for example going over a topic again in more detail or perhaps rewarding yourself with an unexpected break. It is difficult to be precise about how much slack time is needed, but if you planned to revise seven hours in a day, you could leave the seventh hour unallocated to use for whatever was necessary. An example is shown in Figure 13.3.

In addition to planning in slack time you also need to plan in breaks—revising non-stop for seven hours is not an efficient use of time. The longer you go without a break, the more difficult it becomes to concentrate and absorb information. Breaks from revision, therefore, are not a luxury; they are a necessity if you are to be able to revise in a sustainable manner. Therefore, the morning and afternoon slots represented in Figure 13.3 shouldn't be three hours without a break, but rather shorter blocks with breaks built in. Keep your breaks short and free from unhelpful distractions; just getting up to walk round the room or make a drink is enough to renew your concentration. You also need to decide what times of day you work best; some people find working in the mornings easier, others find evenings more productive. Whatever your preference is, try to use it to your advantage, but don't use it as an excuse not to get up in the morning!

Try this—Planning your revision in detail

Using your overall revision plan as a guide, produce a detailed and realistic daily revision plan that you are confident you could stick to.

Do other stuff

Obviously, it is important to be focused on your revision, but it is important to have a life outside of revision, too. The longer your revision lasts the more important it is to factor in time to do other things. Planning in time for other things isn't simply an end in itself, but can be used to help you in your revision in a number of ways, for instance to:

- provide you with a break, so you can come back to your revision mentally refreshed;
- re-energize you, so you can come back to your revision physically refreshed;
- act as a reward for having achieved a certain target in your revision.

Precisely what you do to achieve these benefits will depend on what you find helpful. People often find exercise helpful, or doing something creative, or perhaps spending time socializing. You will need to be disciplined and not get carried away, but a short period of activity, perhaps each day, which you find enjoyable and that will achieve some of the benefits listed will help keep your revision on track and sustain your motivation.

Use variety

In addition to spending short periods of time doing other things, another helpful strategy is simply to use a variety of approaches to your revision. You will know that this is helpful if you have ever tried to spend a revision session just reading, for example. Spending a large amount of time only using one approach can quickly become tedious and demotivating. However, this isn't necessarily a sign that you need a break, it might just be a sign that you need to change your method. Try testing yourself with some questions or flash cards, or drawing a diagram, or writing a summary of what you have just read. We will deal with these, and other, suggestions in more detail shortly when we consider active revision techniques, but the important point to note here is the principle that using a variety of approaches to how you revise can be helpful. A change is often as good as a break (and better than a break if you are using breaks to avoid getting down to revision!)

Go public

Finally in this section on allocating your time, consider making your revision timetable public; you could stick it up on a kitchen cupboard or on your bedroom door. There are a number of reasons this can be helpful:

- it reminds you of what your plans are;
- it informs others who share your accommodation what your plans are;
- it makes clear to others when you are available and when you are not;
- it makes you accountable to others ('I thought you were supposed to be revising?' or 'Shouldn't you be taking a break now?').

Whether or not you want to make your plan public will depend on who you live with, but it is definitely worth considering.

13.1.5 Check that your environment is suitable

Finally, in this section on getting yourself organized, you need to think about the environment in which you will revise. You might be the sort of person who always prefers to work in the same place every time, or perhaps you like variety. Some people find the campus library a helpful place to revise, whereas others prefer to study in their own room. Whatever your preferences, try to make decisions based on what will help your revision to be most effective. A suitable studying environment is one which is free from distractions. For example, it should be:

- well lit;
- quiet;
- not too hot or too cold;
- comfortable (but not to the extent that you fall asleep!);
- reasonably tidy and free from irrelevant clutter.

Sort these things out quickly before your revision session begins, but don't use them as an excuse to avoid starting!

13.2 Use active revision techniques

We said at the beginning of this chapter that you will know when your revision is being effective because it will make your material easier to understand, remember, and apply; this is what active forms of revision help you to do. *Active* revision is any form of revision that makes you interact with the material in an involved and thoughtful way; this includes condensing your notes, drawing summary diagrams, and testing yourself with questions or flash cards. *Passive* forms of revision, however, such as reading or copying, lack the focus provided by active forms and are more difficult to sustain for longer periods.

So, now you're organized, what are you actually going to do when you sit down for your first revision session? Assuming that you have planned your revision timetable so you have a reasonably clear idea of the topics you will be covering on any given day and how long your revision session for the day concerned will last, you are now in a position to begin to work on the detail. This will involve:

- filling in any gaps in your material;
- condensing your notes carefully;
- reviewing your notes regularly.

13.2.1 Fill in any gaps

The starting point, then, is to fill in any gaps in your material that you have identified (as described in Section 13.1.2, 'Check you have all the necessary material') and have decided need addressing. Your material could be supplemented from a number of sources. If you have simply missed a lecture, then borrowing a friend's lecture notes may be adequate, but it is more likely that you will need to do some additional reading. This needs to be done in a certain sequence—make sure you understand the basics first before you progress onto more advanced issues. In practice this means you need to start with lecture notes and handouts, followed by key chapters from core texts on your reading list. If you still need more detail then, once you are confident you understand the basics, you could move onto more specialist publications. This is where it is important that you have understood what you will be examined on (see Section 13.1.1, 'Find out what you will be examined on'), so you are aware how much additional reading is necessary. There will always be more reading that you *could* do; the question you need to answer is whether there is more reading that you *must* do. At this stage, you need to be strategic in your approach—you won't have a lot of time to undertake detailed additional reading; therefore, you need to identify exactly what is necessary and do that, but avoid doing more.

Try this—Identifying gaps in your notes

Find the learning outcomes or objectives of a recent course or module that you have taken. Now go through your notes and identify where in your notes each learning outcome or objective has been addressed. Identify which ones are missing from your notes and identify where you can find the missing information.

13.2.2 Condense your notes carefully

Once you have filled in any gaps in your notes you will probably have a lot of material. Therefore, a vital stage in any revision strategy is to condense your notes into a format that is more manageable. There are two simple reasons for this:

- the process of condensing your notes helps you to learn the material;
- the end product of condensing your notes provides you with a summary of your material, which you can easily review (also enabling you to learn your material).

We will address the review part shortly (Section 13.2.3, 'Review your notes regularly'), but first we will deal with the condensing element. Condensing your notes is an example of an active form of revision—it forces you to make decisions about which parts of your material are the most important and will help you to check how much you understand. This process consolidates your knowledge, and helps you to make connections between different aspects of your learning and identify underlying principles. So, how do you actually condense your notes? Condensing notes involves the following stages (and is illustrated in Figure 13.4).

- taking your original notes on a topic (from various sources) and making a condensed version, perhaps on several sheets of paper;
- then taking this condensed version of your notes and trying to condense the information still further, perhaps to a single sheet of paper;
- finally, writing an overall summary of the particular topic you are revising, perhaps on an index card.

Figure 13.4 Condensing notes

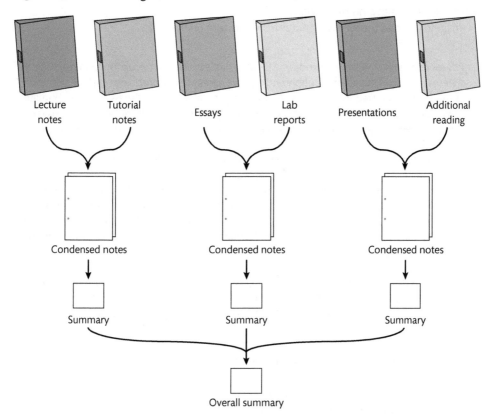

The principle of condensing is simple enough, but how do you decide what information is sufficiently important for you to record and what you can leave out? Clearly, you don't want to simply copy everything out (this would be a passive form of revision, which would quickly become ineffective), but equally, you don't want to be so sketchy with your notes that when you come back to them at a later stage there is insufficient detail for you to work from. This is why this process involves a number of stages—you can't skip from your lecture notes straight to an overall summary without going through the intervening stages. The reason for this is that, in producing condensed notes and then summaries of these condensed notes, you are not only producing a more condensed version of your material on paper, but also structuring and connecting the information in your mind, and making decisions about what is sufficiently important to record and what you can leave out. At each stage you are attempting to distil out the essence of the particular topic you are studying, in order to understand it better. What you will be left with, therefore, is not everything you need to know on the topic, but sufficient information to prompt you to recall the source information from which it was distilled.

Exactly how much information you record depends on how much you already know and so will vary from individual to individual. What you are aiming for, however, is to pick out key facts and create structures to help you to understand whole concepts. Thanks to Professor

Stewart Petersen from the University of Leicester Medical School for the following example, which illustrates how this might work.

Imagine you had the following notes from a lecture on hypoxia.

Tissues consume oxygen and produce CO_2 at variable rates. The circulatory and respiratory systems must work together to supply and remove gases at an appropriate rate for each tissue of the body. Each step in the chain of supply can be affected by disease, but the ultimate consequence of any such condition is that some or all parts of the body receive less oxygen than they need—hypoxia, sometimes also associated with the inappropriate accumulation of carbon dioxide.

The supply of oxygen to each individual tissue depends, first, on appropriate oxygen content in arterial blood, and second, on adequate local perfusion.

Appropriate oxygen content in arterial blood depends on effective diffusion of oxygen from alveoli into alveoli capillaries, and appropriate partial pressures of oxygen in the alveolar gas.

Appropriate pO_2 in alveolar gas depends upon adequate ventilation of the alveoli, which itself depends upon the ease of air flow through the airways, the capacity of breathing movements to generate appropriate pressure differences, and appropriate composition of inspired gas.

Any disease process affecting oxygen supply to tissues must affect one or more of these steps, which therefore provide a convenient classification for pathophysiology, and the ultimate outcomes, local or generalized, are poor supply of oxygen—hypoxia.

A useful way to start to summarize such notes is to annotate them as you read them, as shown in Figure 13.5.

You could also choose to summarize the content in your own words, as shown in Figure 13.6.

To condense the information further you could create a summary diagram, as shown in Figure 13.7. This summary diagram has the added advantage of including additional information.

In addition to considering the content of the information you are condensing, you also need to consider its format. The example used in Figure 13.7 is a deliberately visual example. This is because many people find it easier to remember something that is visual, rather than something that is essentially text-based. In Chapter 3, 'Making the most of lectures', we suggested a number of points to improve your note-making; many of these are also relevant for note-making for revision. Remember to try to use:

- headings and sub-headings;
- colour;
- space;
- figures, including labelled diagrams and flow charts.

Try this—Producing a summary diagram

Think about the last lecture that you attended. Look at the lecture notes or handout for that lecture, and summarize them into a single page summary. Now produce a summary diagram based on that lecture.

Figure 13.5 Annotated notes

Hypoxia

Tissues consume oxygen and produce CO_2 at variable rates. The circulatory and respiratory systems must work together to supply and remove gases at an appropriate rate for each tissue of the body. Each step in the chain of supply can be affected by disease, but the ultimate consequence of any such condition is that some or all parts of the body receive less oxygen than they need – **hypoxia**, sometimes also associated with the inappropriate accumulation of carbon dioxide.

The supply of oxygen to each individual tissue depends first on appropriate oxygen content in arterial blood, and second on adequate local perfusion. Appropriate oxygen content in arterial blood depends on effective diffusion of oxygen from alveoli into alveoli capillaries, and appropriate partial pressures of oxygen in the alveolar gas.

Appropriate PO_2 in alveolar gas depends upon adequate ventilation of the alveoli, which itself depends upon the ease of air flow through the airways, the capacity of breathing movements to generate appropriate pressure differences and appropriate composition of inspired gas.

Any disease process affecting oxygen supply to tissues must affect one or more of these steps, which therefore provides a convenient classification for pathophysiology, and the ultimate outcomes, local or generalized, are poor supply of oxygen—hypoxia.

Figure 13.6 Summary notes

- tissues consume O_2 and CO_2 @ variable rates (circ. + resp. systems together supply + remove gases @ apt. rate)
- each step of supply can be affected by disease → less O_2 than needed (hypoxia)
- supply of O_2 to tissues depends on
    1. apt. O_2 content of arterial blood
    2. adequate local perfusion
- 1 depends on effective diffusion of O_2 from alveoli → alveoli capillaries & apt. PO_2 of O_2
- which depends on adequate ventilation of alveoli
- which depends on ease of air flow thro' airways, breathing movements + comp. of insp. gas

NB. diseases affecting O_2 supply to tissues must affect at least 1 step

Figure 13.7 Summary diagram

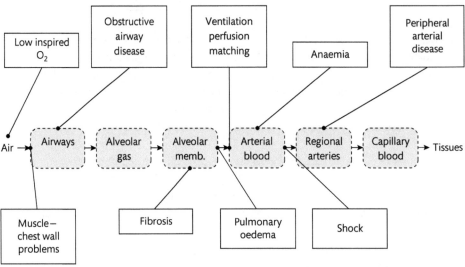

13.2.3 **Review your notes regularly**

We observed earlier that condensing your notes provides you with a summary that you can easily review. If you have ever tried to review uncondensed notes, you will know that it is a very difficult exercise to sustain because of the sheer volume of information you are trying to read. Even if you manage to review uncondensed notes for a particular topic, the more topics you have the more unsustainable it becomes. If you have produced summaries, however, not only do you have your notes in a handy format for you to review, as identified in Section 13.2.2, 'Condense your notes carefully', but the process of condensing your material will also have helped you to learn it. Clearly, you won't remember everything (that is why the review stage is important), but you will remember some and you will also have something that is reviewable.

The importance of reviewing your notes regularly can be highlighted by considering what happens when you don't do so. What inevitably happens is that the longer you go without reviewing your material the more you forget. Without regular reviews of what has been revised you won't retain the information you have taken the trouble to learn.

The relationship between reviewing and recalling information is straightforward enough, but how do you actually review your material? There are many different methods of reviewing information, but whichever method you choose, it needs to test your recall of a topic by:

- checking your ability to understand central concepts; or
- remembering key facts; or
- linking the information to other topic areas; or
- perhaps all three!

As we have already noted, you won't remember absolutely everything, and you can always go back to your original notes to remind yourself of particular details; the important thing is to do the reviewing. Methods of review include:

- testing your recall of key facts with index cards (or flash cards), putting to one side the cards you can remember and repeating the cards that you can't;
- reproducing from memory headings and keywords for a learnt topic area and then checking them against your notes;
- recording yourself explaining concepts or theories to an imagined audience, then playing back your explanation, checking it against your notes;
- making a visual summary (like the flow chart in Figure 13.7) to summarize information you have remembered;
- reproducing important diagrams that summarize large amounts of information or whole concepts, e.g. metabolic pathways;
- enlisting the help of a friend to test your recall using any of these methods;
- putting up summaries of key information around your home to give you the opportunity for on-the-spot tests;
- testing yourself with question banks and past papers (more on this shortly).

Reviewing your material need not take a lot of time. A good time to review material is at the beginning of a revision session, but it can also be fitted into any spare time during the day. Frequently reviewing your material will help to build up your confidence and reassure you that you are making progress.

13.3 See the big picture

We have noted in Section 13.2, 'Use active revision techniques', that condensing your notes supports the consolidation of your knowledge, and helps you to make connections between different aspects of your learning and to identify underlying principles. However, when paying attention to detail, it is easy to lose sight of the context and principles. It is important, therefore, that you make a special effort to appreciate the big picture. Try not to get so focused on the details that you fail to see either where the information fits into the larger context or perhaps a simple principle that underpins it, both of which would make it easier to understand and remember.

The exact parameters of the big picture will depend on how much breadth and depth you are required to know, and this will change depending on the module choices you make throughout your course and the level of study you are at. However, context and principles are important to appreciate and identify because they help you to develop a means of structuring, storing, and retrieving information. They also help because exam questions rarely ask the precise question you had been hoping for; rather they require you to apply your knowledge to different situations.

Imagine walking into a room with papers scattered across the floor and then being asked to find a particular piece of information—it would be very difficult! If you did manage eventually to find the correct piece of paper it would be difficult to then explain how the information on the piece of paper relates to all the other information on the pieces of paper still on the floor. However, imagine the same papers in the same room but this time organized appropriately into a filing cabinet. Suddenly it becomes much easier to find the required information and relate it to other information. Seeing the big picture is a bit like storing papers appropriately in a filing cabinet (or information in a database); it enables you to appreciate how information fits into a wider context and to identify the principles that underpin it. Without these elements the information you are trying to learn will remain abstract and unrelated, which makes it much more difficult to understand, remember, and apply. It is important, therefore, that you spend some time standing back from the detail to think about structure. Appreciating context and identifying principles are two specific ways of doing this, but anything that helps you to reflect on and question the information you are revising, helping you to understand themes and relationships, rather than detail, will help you to see the big picture.

13.4 Practise outputting the information you have learnt

We have noted the importance of understanding, remembering, and applying the information you learn. However, the steps we have focused on so far are largely concerned with the first two elements: understanding and remembering. If you are going to perform well in the exam room (more on this in the Chapter 14, 'Getting the most out of exams'), then you need to be able not only to understand and remember information, but importantly, you need also to be able to apply it. This is because exam questions, whatever the format, are specific and you are, therefore, required to give specific answers. When you open an exam paper you are very unlikely to find a question that reads 'Please tell us everything you know about comparative animal physiology' (although it's surprising how many answers are written as if this was the kind of question asked!). A much more likely question would be 'Compare and contrast the retinal organization of a named mammal and reptile.' Therefore, the information you have taken time to understand and remember needs to be applied to the specific question. It would be very unfortunate if you had taken the time and effort to understand and remember information, but failed at this last crucial step. The best way to improve your ability to answer questions specifically is to practise. The two main ways of practising are:

- creating question banks; and
- using past papers.

13.4.1 Create question banks

A question bank is simply a collection of questions that you think you might be examined on. The questions shouldn't be full essay questions, but short, one-line questions that test very specific knowledge. Question banks are useful for a number of reasons:

- past papers may not always be available to you;
- the format of the exam might have changed;

- thinking up questions helps you to reflect on how much you know;
- answering the questions helps you to apply your knowledge.

Question banks can be created from any questions that you think you might need to know the answer to. A good place to start is the learning outcomes for the modules—use these as a prompt to help you think of questions. Keep adding questions to your question bank throughout your course; questions prompted from lectures, or tutorials, or reading. You can then use these questions to check your knowledge of the material and so monitor your progress, or as a means of review, or to supplement your use of past papers. It is also beneficial to team up with some fellow students to share the task of creating a question bank; this will save you time as individuals and give you a greater variety of questions than you might have created on your own.

Try this—Creating a question bank

Get together a small group of your fellow students on a course you are revising for. Ask each of them to come up with three questions that they think might be important to be able to answer, based on the learning outcomes or objectives of the course or module. Share these questions between you to form a substantial question bank.

13.4.2 Use past papers

Past papers are an extremely helpful resource when you come to practise outputting the information you have learnt. If you have already found out what you will be examined on (Section 13.1.1, 'Find out what you will be examined on'), you will know what the format and scope of the exam will be. Assuming there are no major changes to an exam's format or scope for that particular year, you can assume that the past papers will be representative and so be a good test of your learning so far. There are a number of ways you can use past papers, including:

- practise analysing questions and setting up plans to answers in order to test your recall and ability to adapt material to the set question;
- practise writing full answers, by hand (i.e. not typed), in the time allowed—as most exams are still hand-written this will also help you practise writing legibly over an extended period and under time pressure;
- review your answers by checking them against your notes and highlighting any missed or inaccurate information;
- ask one of your lecturers to read your essay and comment on it;
- analyse your answer to see how it could have been improved. (Did it need more information? Did it have a logical structure? Was your expression clear? Did you stick to the question?)

Your precise use of past papers will depend, in part, on the format of the exam: multiple-choice, short-answer, or essay. We will deal with this in more detail in the Chapter 14, 'Getting the most out of exams'.

Chapter summary

We said at the beginning of this chapter that effective revision makes the material you are revising easier to understand, remember, and apply. Getting yourself organized in your revision, with a clear timeline leading up to the exam, is a crucial first step in revising effectively and can reduce the amount of stress that exams cause. Making your revision an active, rather than passive, process will make your revision both more interesting and more effective. Making sure that you don't lose sight of the big picture in terms of how detail fits into an overall structure is also important, as is practising outputting the information you have been learning to ensure you can apply it to specific exam questions.

14 Getting the most out of exams

 Introduction

We ended Chapter 13 on 'Getting the most out of revision' by addressing the use of question banks and past papers as a means of practising how to output the information you have learnt. This chapter on 'Getting the most out of exams' will address the skills you need for the exams themselves. However, exam skills are not just needed during exams; there are some important exam skills you need to develop for the day or so leading up to exams, too. With this in mind, this chapter will begin by addressing what you need to do immediately before your exams, and then progress to give guidance about the exams themselves.

Remember: unlike most other topics covered earlier, you won't have this book to hand for guidance when you're actually experiencing the topic of the chapter—that is, when you are sitting in the exam room. So it is worth spending time reading carefully to make sure you remember all the key points, so that you have them in mind and ready to put into action when exam time comes.

14.1 Just before an exam

If you have got yourself organized and managed to do a reasonable job of your revision, as described in Chapter 13, 'Getting the most out of revision', then there will be less of a temptation to do any last-minute cramming. You need to be realistic at this point, however—there will always be more you could do, but at some point you need to stop. It is very unusual, even for students who have managed to get themselves very organized, to be able to walk into an exam room feeling completely prepared. (After all, exams are—by their very nature—an unknown quantity; the only way to be 100% prepared would be to know the questions before the exam!) It is, therefore, common to want to try to cram in a lot of revision just before an exam (either because your revision hasn't gone as well as you had hoped or because you are anxious), but if you do find yourself in this situation, then be careful! There is work that is helpful to do at the last minute, but equally, some which can be unhelpful.

14.1.1 **The wrong sort of last-minute work**

Last-minute cramming is the wrong sort of last-minute work. There are a number of reasons for this.

Last-minute cramming is ineffective

If there is material that you still haven't covered the night before the exam you are unlikely to gain much benefit by covering it at this late stage. Your time would be much better spent going over things that you are already familiar with (see Section 14.1.2, 'The right sort of last-minute work').

Last-minute cramming makes you forget other stuff

Last-minute cramming is often counterproductive, because trying to cram in a lot of information (particularly new information) at the last minute can cloud your memory of what you already know and so has the effect of pushing out other information that you may have otherwise been able to retain.

Last-minute cramming makes you tired

Exams are hard work, and require a lot of energy and concentration. Therefore, making yourself tired by staying up late the night before (or worse, through the night) is not good preparation. Your time would be much better spent trying to get some rest so that you are alert for the exam and can make a good attempt at answering the questions, rather than being so tired you can't think clearly.

Last-minute cramming makes you worry

Last-minute cramming tends to make you focus on what you don't know, rather than what you do. Thinking about all the things you don't know the night before an exam is a bad idea! It makes you worry, it gets you stressed, it makes it difficult for you to get to sleep, which makes you more worried and more stressed . . . and so it goes on.

14.1.2 The right sort of last-minute work

While last-minute cramming is the wrong sort of last-minute work, there is work you can do just before an exam which is helpful.

Plan your use of time

It is important to plan how you will use the time you have in the exam *before* you get to the exam room. This doesn't have to be a last-minute exercise, but if you haven't done it as part of your revision strategy, now is the time to do it. You need several pieces of information to plan your use of time effectively:

- how long the exam lasts;
- how many questions the paper contains;
- how many of the questions you are required to answer;
- whether all the questions are worth the same number of marks.

Once you know this information you can decide how much time you will spend on each question, remembering to allocate time for choosing questions (if you have a choice), analysing questions, planning answers, and reviewing answers, as well as (of course) time to write your answers. For example, if you had a one-hour short-answer paper containing nine questions, each of which was worth ten marks, you might work out how to allocate your use of time as follows:

- one hour = 60 minutes;
- five minutes per question (including analysing and planning time) = 45 minutes;
- which leaves 15 minutes checking time;
- 15 minutes checking for nine questions = just over 1.5 minutes checking time per question.

If you had a three-hour essay paper containing ten questions, all worth 25 marks, of which you had to answer three, you might allocate your use of time as shown in Figure 14.1.

Planning your use of time before the exam will give you one less thing to think about in the exam itself. It will also ensure that you at least attempt all the questions you are required to attempt, rather than simply focusing on the ones you are confident on. In an essay paper where you are required to write three essays, for example, you will almost always get more

Figure 14.1 Example plan for use of time for a three-hour essay paper containing ten questions of which you are required to answer three questions.

Time (minutes)	Task
5	Read questions
5	Choose questions
50	Question 1 • analysing (5) • planning (5) • writing (30) • reviewing (10)
50	Question 2 • analysing (5) • planning (5) • writing (30) • reviewing (10)
50	Question 3 • analysing (5) • planning (5) • writing (30) • reviewing (10)
20	Checking

marks by having a go at all three essays, rather than spending time trying to make two of them really good ones. Even if you are not confident about the third essay and think you can only write a fairly poor answer, the chances are you will get more marks for attempting an answer to the third one than by refining an existing one.

The night before the exam

We have already noted that the night before an exam is not a good time for learning new information. It is, however, a good time for practising your exam technique or checking your knowledge of what you have already learnt. So use your question banks and past papers to go over the information you have already covered, making sure that if you come across something that you don't know you look it up so you do know it. Going over pre-revised material in this way the night before an exam will both boost your confidence regarding how much you do know (you might be surprised!) and act as a reminder of the things that you had learnt, but couldn't recall in response to a question.

Do make sure, however, that you don't work late on the night before the exam and try to leave some time for relaxing before you go to bed, as a good night's sleep is important for doing your best the next day.

The morning of the exam

The morning of the exam is also not a good time for learning new information. As we have already mentioned, cramming can make you forget other stuff; in particular, it can cloud your ability to remember overall concepts. Instead, focus on reviewing main points to give you an overview of what you have learnt. If your exam is in the morning, it would be a good idea not to do any sort of revision on the morning of the exam at all.

14.1.3 Other last-minute preparations

Good exam technique is about more than simply learning information and then outputting that information in response to specific questions. Your preparation could have gone brilliantly, but if you are so tired you can't think clearly, or you turn up at the wrong location, your preparations will have been wasted. So here are a few more last-minute preparations to consider.

Get your stuff together

Getting everything you will need for the exam organized the night before will give you one less thing to think about on the day itself. You will need a pen you can write with comfortably for a long time, and at least one spare. Depending on the exam you may also need other equipment, such as a calculator. Also think about what you don't need, which is anything that might cause an invigilator to suspect you of cheating if they found it on you. This includes mobile phones, audio players, scraps of paper with notes on, and scribbles on the back of your hand.

Try this—Preparing for an exam

Several weeks before your next exam find out the date, the time, and venue. Find out how long before the exam starts you will be allowed into the examination hall and what equipment you can take in with you. Make sure you know how many questions are on the paper, how many questions you have to answer, and how long you have to answer each one.

Double check when and where you need to be

Another thing you don't want to be worrying about on the morning of the exam is when and where the exam will take place. Make sure you have the latest information on the date, time, and location of your exam (sometimes these change due to timetable clashes, which is why the *latest* information is important). Then double-check this with other people who are taking the same exam, just to make sure. If your exam is taking place in a location you are not familiar with, it is worth either going to the venue beforehand or arranging to go with someone on the day who is sure they know where it is. Also, set your alarm to make sure you wake up in time, and then set another alarm, just in case. You may also want to let the people you live with know when you need to be up, so they can help make sure you're up, too. Having more than one wake-up call reduces the chances of you worrying about not waking up in time and so makes it more likely that you will get to sleep. Plan to arrive at the exam room in plenty of time, as racing in at the last minute (or late) flustered and out of breath is not a good way to start.

Get some sleep

We have already identified that last-minute cramming isn't helpful, but a good night's rest the night before the exam is. However, this is easier said than done, so it helps to have a strategy. First, make sure that you don't drink too much caffeine too late, as clearly this will keep you awake. Secondly, it is very unlikely that if you stop working at 11 p.m. and then roll straight into bed you will fall asleep quickly, and the longer it takes you to fall asleep the more worried you will become about not being able to sleep and so the more awake you will be.

To help you relax and sleep well plan something relaxing to do in between stopping work and going to bed. It could be chatting to your housemates, going for a walk round the block, or watching a bit of television (as long as you don't start a whole film!). Anything that will help stop your mind racing and calm you down a little will make it more likely that you will be able to sleep.

Manage your stress levels

We have deliberately given this section the title 'Manage your stress levels', rather than 'Get rid of stress' because, first, it's unlikely that you will be able to get rid of stress altogether, and, secondly, some stress is helpful—it is your body's way of preparing itself to cope, making you more alert and attentive. All that we have covered so far in this chapter and Chapter 13, 'Getting the most out of exams', will help you to manage your stress levels, because one of the most stressful things, as far as exams are concerned, is lack of preparation. So getting organized will really help—the most stressful exams are always the ones you are not organized for. If you still feel anxious though, it can also be useful to talk to someone about your worries concerning exams. This could be a friend, your tutor, or someone in your university's student support services. Additionally, on the day of the exam itself, there are a number of things that will help you manage your stress levels:

- *try to avoid talking to other students* if you think it will increase your anxiety levels;
- *get plenty of fresh air*—exam rooms are often stuffy places, so get plenty of fresh air beforehand (this will also help you avoid talking to other students), and breathe deeply to try to calm your nerves;
- *make sure you use the toilet before the exam*—you really don't want to be needing to go halfway through the exam—it will be a major distraction, and will make you feel unsettled (and, therefore, stressed).

Exams are physically tiring. So there are also a couple of things not directly related to the management of stress that you should keep in mind to make yourself as prepared as possible from a physical point of view.

- *Try to eat breakfast*—not everyone finds this easy, but exams are hard work and you need energy. At the very least, have something simple, but high-energy, like a banana.
- *Drink water*—to keep you hydrated and help you think clearly (and take a bottle of water in with you).

14.2 Arriving in the exam room

When you arrive in the exam room there are several things you can do to get yourself off to a good start.

14.2.1 Get settled

Exams have a lot of build-up and there is often a lot depending on them. So it is inevitable that you will be feeling at least a little anxious. Just by sitting down, taking a few deep breaths, getting out your pens, your clock or watch, and your bottle of water, and generally getting yourself comfortable in your chair will help you settle down.

14.2.2 Read the instructions very carefully

Once you have been told by an invigilator that you can turn over the paper, turn it over. The first thing to check is that it is the paper you were expecting (if it isn't, put your hand up—often exam rooms contain several different exams at once; you could just be at the wrong desk). Once you are sure you have the right paper, begin to read the instructions carefully. The instructions should be familiar to you from your preparation, but check in particular the number of questions, how many questions you are required to answer, and how many marks each question is worth.

14.2.3 Remind yourself of your time plan

You should have planned your use of time beforehand (see Section 14.1.2, 'The right sort of last-minute work'), so this should simply be a case of reminding yourself what you had decided to do (but if you haven't—do it now). You may find it helpful to make a note of your time allocation on the question paper, so you don't forget or become confused about what you had planned to do (which is easily done when you are under pressure).

14.3 Answering the questions

The areas we have covered so far in Sections 14.1, 'Just before an exam' and 14.2, 'Arriving in the exam room' apply equally to all types of written exam, whether multiple-choice papers, short-answer papers, or essay papers. However, when it comes actually to answering the questions there are some exam techniques that are particular to different types of written exam. We will deal first with some general advice applicable to all question types, and then address how these apply specifically to different types of exam paper.

14.3.1 General advice for all question types

Regardless of the type of exam paper you are faced with, it's a good idea to have a clear sequence of steps in mind to help you answer the questions to the best of your ability. We suggest the following:

- analyse the questions;
- plan your answers;
- write your answers;
- review your answers.

Following these four steps will help ensure that you have a measured approach to what can otherwise be quite a panicky situation.

14.3.2 Multiple-choice papers

Of the question types we have identified, multiple-choice papers require the least amount of analysis and planning. However, that's not to say you can dispense with analysis and planning altogether—you just need to approach it in a slightly different and briefer way.

Analyse the questions

Usually, with multiple-choice papers, you will not have a choice of questions you can answer. If this is the case, then there is no need to read through the entire paper before you begin. Instead, simply read the instructions carefully (as described in Section 14.2.2, 'Read the instructions very carefully') and make sure that the structure of the paper is what you expected. Once you have confirmed this, you can begin to work on the individual questions. With multiple-choice papers you should normally start with the first question and then work through the questions in sequence. When you come across a question you don't know the answer to, don't dwell on it too long, just move on to the next one, and come back to the ones you are not sure about at the end.

Plan your answers

You neither have, nor need, much time to plan your answers for multiple-choice papers. However, a few moments spent planning will help make sure you don't rush headlong into a wrong answer. A good approach is as follows:

- *read the question carefully*;
- *don't look at the possible alternatives yet*—try to answer the question independently of the options given;
- *look down the list of options* to see if your answer is there;
- *read all the other alternatives* just to make sure.

Write your answers

For multiple-choice questions this is simply a case of ringing a response or ticking a box, but do make sure that you have read the question carefully. Sometimes multiple-choice papers include multiple-response questions, as well as multiple-choice questions, that is to say questions that require more than one answer. For example:

Which of the following statements about the neuron is true?

- The resting potential is dependent on the membrane permeability to Na^+ ions ☐
- The resting potential lies close to the equilibrium potential for K^+ ions ☐
- The membrane is only permeable to Cl^- ions ☐
- The membrane is impermeable to K^+ ions ☐

In this question, as indicated by the stem: 'Which . . . is *true*?', there is only one correct answer, (b).

In some questions, there may be more than one true statement. These are indicated in two ways. For example:

Which of the following statements about the lungs are true (tick all that apply)? . . .

Or some take a more complicated form, as follows:

Which of the following statements about the ventilation of the lungs are true?

(a) The expansion of the lungs is mainly as a result of contraction of the diaphragm.

(Continued)

(b) The expansion of the lungs has to overcome the resistance due to surfactant.

(c) Expiration at rest is the result of contraction of the intercostal muscles.

(d) The tidal volume is typically 500 mL.

(a), (b), and (c) are true	☐
(a), (b), and (d) are true	☐
(b) and (c) are true	☐
None of the above	☐

This second type of question is quite complicated and you need to work through the statements carefully, marking each one that you think is true and then seeing how your list matches with the options given.

Also, make sure you have understood the sense of the question. For example, check whether the question is asking you to indicate which is true or which is *not* true.

Your exam questions may also be set out as Extended Matching Sets (EMS). A simple version of an EMS question is:

1. During quiet respiration, the main contributor to inspiration is contraction of the .... During forced expiration the passive relaxation is supplemented by contraction of the .... and the .... muscles.

2. In actively metabolizing tissues, the pH is .... and the oxygen dissociation curve shifts to the right, leading to .... affinity. This is termed the ....

3. During heavy exercise, the arterial pCO_2 may be .... as a result of the ventilatory response to the production of ....

4. Taking several deep breaths prior to diving extends underwater endurance because the .... of the blood is initially.....................

External intercostal	Decreased
Pectoral	pH
Bicarbonate	Lactic acid
Increased	Diaphragm
pO_2	Internal intercostal
Bohr shift	Chloride shift
Abdominal	Tidal volume
Unaltered	Buffered

Here, you have been given a number of incomplete statements and a table of words that could fit those statements. You have to complete the statement with the most appropriate word, bearing in mind that each word may be used once, more than once, or not at all. Again, the best strategy is to read through the statement and think what should go in the space first, then look at the list and see if any of the words match. That way you are less likely to be confused by the distracters.

Review your answers

When you get to the end of the paper, check back through the paper and re-read those questions that you were not originally sure about, making sure that you have answered all the questions. The only time when you should change this strategy is if you have a paper where the questions have negative marking (this type of paper is not very common). In this case, incorrect answers will actually cost you marks, so you may be better off leaving some questions blank if you are really unsure of the answer. After having checked for uncompleted questions, review your answers, starting with the ones you were least sure about. If you amend an answer make your change very clear; there will be advice on how to do this in the instructions at the beginning of the paper.

14.3.3 Short-answer papers

Short-answer papers can also be tackled using the same four steps. Again, one of the keys to doing well on these papers is being strict with your timing—don't spend a lot of time thinking about each question. If you know the answer, answer it quickly, if you don't know the answer, move on and come back when at the end, when you may have time to think about it more carefully.

Analyse the questions

As with multiple-choice papers, short-answer papers often don't give you a choice of which questions you can answer. Again, if this is the case you don't have to read through the entire paper before you begin. Start with the first question and read it carefully to make sure you understand it, perhaps underlining key words to help you focus on the exact meaning. If after a quick reading you are not sure how to answer it, move on to the next question and come back to the ones you are less sure of at the end.

Short-answer questions usually require specific factual information for the answer, and not detailed description. Usually, the number of marks to be awarded is given on the exam paper, so that can be a guide to the amount of information you need to give. Typical short-answer questions might take the following forms:

---

1. Draw a labelled diagram of the nerve action potential. (4 marks)
2. Label structures A–D on the diagram of the ear. (4 marks)
3. List the stages involved in the replication of DNA. (5 marks)
4. Describe the process of synaptic transmission. (5 marks)

Plan your answers

With short-answer questions, in briefly planning your answers you are trying to make sure that you select information relevant to the question, and put it into a logical and coherent order. In each of the questions in the box, the marker is looking for very specific factual information. To answer the question, you will need to note down a few keywords to stimulate your memory and organize your thoughts.

Write your answers

Use your brief plan to prompt you to make your points and structure your answer. You only have a short time to write each answer, so the important point is to make sure that all the key facts are included. For example, the diagram of the nerve action potential does not need to be a great work of art, but you must remember to include key details such as scales to mark the voltage and the time, as well as the actual factual content. When writing your answers, brevity and factual content are the keys to a good answer. It is also important that you write legibly—this can be difficult if you are in a hurry and if you are not used to hand-writing for long periods (which is why we said in Chapter 13, 'Getting the most out of revision' that it is important to practise writing by hand for extended periods). Take some time to make sure your handwriting is legible.

Review your answers

Check that you have answered every question that you needed to answer (probably all of them), then review your answers. As with multiple-choice papers, start with the questions you are least sure about because these are probably the ones you will be able to improve the most.

14.3.4 **Essay papers**

Often with essay papers you will have a choice as to which questions to answer. If this is the case, then make sure you read through all the questions carefully before you choose which one (or, more likely, ones) you are going to answer. As you read the questions, mark the ones that you think you will be able to answer the best, then go back and read them again to make a final choice. As with multiple choice and short answer questions, start with the questions you are most confident with to get you off to a good start and stimulate your thinking.

Analyse the questions

Analysing questions is particularly important for essay questions. With multiple-choice and short-answer papers your answers will usually just be facts or isolated pieces of information. With essay questions, however, there is more work to do to at the analysis stage in order to be able to organize your answers logically and coherently. As we said in Chapter 13, 'Getting the most out of revision', essay questions will usually be very specific. So, instead of asking 'Please tell us everything you know about comparative animal physiology', they are much more likely to ask you to, for instance, 'Compare and contrast the retinal organization of a named mammal and reptile.' Applying your knowledge to this particular question requires you to analyse the question briefly, using the same technique that we identified in section 7.3.2, 'Analyse the question or title'. The key information to identify in the question or title is:

- the subject of the question;
- the instruction;
- the key aspect;
- other significant words.

Plan your answers

Planning answers is particularly important for essay questions because it helps you to:

- select information that is relevant to the question;
- put this information into an order that is logical and coherent;
- write your ideas down at an early stage to help you remember your key points;
- monitor how much information you are covering in the time allowed.

Write your answers

When answering essays under exam conditions it is important that you keep to your time plan, as it can be easy to lose track of time if you get into the flow of writing. One way to do this is to make a point of checking the time when you come to the end of a section in your answer. This is preferable to checking the time at random intervals (which could be insufficient) or too frequently (which could distract you from writing fluently). Also, keep looking at the question to remind yourself of the focus of your answer, and keep looking at your plan to remind yourself of what you need to cover. Again, the end of each section of your answer makes a good natural break for checking that you're still on track. Make sure that, as you write, particularly as you get into the flow of writing, that you keep your handwriting legible.

Review your answers

Being able to review your answers is where the benefits of having planned and monitored your use of time become obvious. Not only should you have had enough time to make a reasonable attempt at each of your chosen questions, you should also have time left at the end to go back and review your responses. As we highlighted in Chapter 7, 'Writing essays', making time to review your work can make a significant difference to your final grade.

As with multiple-choice and short-answer questions, first check that you have answered every question that you need to answer, then review your answers, starting with the one you were least sure about. The purpose of reviewing answers in an exam context is slightly different to reviewing answers in a coursework setting—you don't need to finish up with a very polished piece of writing, which should be spelling mistake-free and grammatically faultless, just one that answers the question well and is legible. Of course, if you do end up with a polished piece of writing then that's good, but it's not the primary purpose of a review in an exam context.

Instead, you should be looking at the bigger issues, such as:

- Have you answered the question?
- Does the order of your material make sense?
- Is the meaning of what you've written clear?
- Have you made appropriate use of figures to supplement what you've written?

These are the main issues; you will get more marks for a response that has spelling and grammatical errors, but answers the question well than you will for a response that has perfect spelling and grammar, but doesn't answer the question well.

If you need to make significant structural changes to your answer, such as moving a paragraph from one section to another, you need to do this in a way that does not confuse the people marking it. In a word-processed script this would be straightforward, but in a handwritten script you will need to annotate it in a way that makes it clear you are moving a section, rather than, for instance, deleting it.

 Chapter summary

Performing well in exams requires particular skills. Although your exam performance will be heavily influenced by how well you have revised, it is how you conduct yourself during an exam that will ultimately determine your grade. What you do just before an exam and when you arrive in the exam room are important stages of your preparation. Different types of question require slightly different techniques, but all require you to analyse questions, and plan, write, and review answers.

Making yourself employable

 Introduction

We considered putting this chapter at the beginning because thinking about it at the end of your degree is far too late. However, we identified making yourself employable in the opening chapter as a foundational skill, and have highlighted the employability applications of study and communication skills at relevant points, so hopefully you have got the message that making yourself employable is vital! Having said that, if you are reading this chapter in your final year do not despair! This chapter can still help you to make the most of the time you have remaining at university to make yourself employable. The wording of the title is important, too—I'm afraid we can't guarantee that following the advice in this book will get you into *employment* (though we certainly hope it will), but we are confident that if you follow the advice it will make you *employable*. And when you think about it, being employable is, in many ways, more important than being employed—jobs no longer last a lifetime, and when you find yourself looking for work it's employability that counts. Peter Hawkins in his book *The Art of Building Windmills* (1999, Graduate into Employment Unit) put it like this: 'To be employed is to be at risk, to be employable is to be secure.' So regardless of what the current job market happens to be doing, the important thing is to make yourself employable, so that you can increase your chances of being employed whatever the circumstances. This chapter will highlight the importance of getting a good degree, engaging early with employability, gaining experience, making contacts, and selling yourself in the context of the selection process.

15.1 Get a good degree

This book is all about helping you to get a good degree, and if you've spent a lot of time, effort, and money on your studies you'll want to get a good result. The classification of your degree is one of the first things that many employers will look at in their selection processes. The larger graduate employers may receive hundreds of applications for each position so they will want to filter out applicants based on easily quantifiable criteria, such as degree classification and even UCAS tariff points. This means, that for many graduate recruiters (but importantly not all), if you don't get a 2:1 degree you won't get past the application form stage. Certainly, all is not lost if you get a 2:2, but it will make your choices more limited. If you get a 2:2 (or think you are heading for one) you will have to research which employers are happy to accept a 2:2 or don't use degree classification in their screening process (there are a growing number of these and your university careers service will be able to help you with this) and then focus on those. Bear in mind, however, that some employers who have a 2:1 as a minimum criterion may choose to ring-fence a handful of places for graduates who have a 2:2, but can demonstrate outstanding strengths in other areas (e.g. skills, experience, or attributes). Having said all that—it's much better to try to get a 2:1 (or 1st) in the first place!

15.2 Engage early

As we said in the introduction, trying to make yourself employable at the end of your degree is far too late; you have to engage early. Admittedly, thinking about your CV is probably not at the forefront of many first years' minds during freshers' week, but once you've had a couple of weeks to get settled it's a good idea to begin to think about what you want to do when you graduate, however far off it may seem. The reason you need to start early is because a good degree is not the only thing that employers look for. In fact, in many ways, employers take the degree as a 'given' and it's everything else that really counts. This is less the case if you are using your degree to get into an industry that requires very specific academic qualifications, but regardless of which industry you are targeting it's always the additional things that make you stand out. Employers will be looking for a good degree plus:

- transferable skills (team working, communication, project planning, problem solving, etc.);
- relevant experience (dealing with customers, positions of responsibility, etc.);
- personal attributes (motivated, down-to-earth, get on with others easily, etc.);

and all of these will need to be effectively communicated through the appropriate part of the selection process, whether that be by CV and covering letter, application form, assessment centre, or interview.

A useful maxim is this: 'Figure out where you want to be and then work backwards'. That is to say, what is it you want to do and what do you need in order to get there? For example, if you want to be a teacher you will need experience of working with children and to have had a number of placements days in a school. Couple this with the fact that applications for teacher training need to be submitted 6–12 months in advance (depending on which route you choose) and you can see why you need to start early. For teaching, ideally, you need to begin working on your relevant experience early in your penultimate year if you want to get on a teacher training course when you graduate. Likewise, if you want to use your degree to work in the pharmaceutical industry, then you would be advised both to aim for a practically based project and to obtain some additional laboratory experience during your undergraduate studies. This experience could take the form of a placement in a university or other research lab during the vacations, or as a sandwich year with a placement in industry. So, if you want to follow this route, you need to be planning early, probably at the start of your second year. You also need to choose your final-year project carefully, aiming for one that will give you more relevant laboratory experience.

Saying that you should 'figure out where you want to be and then work backwards' is all very well if you know what you want to do (or at least have a rough idea). You may not know what you want to do though, in which case you need to start a bit further back.

15.2.1 Know yourself

There are many good books and websites available on career planning, so we are not trying to cover everything here—just the basics. First, you need to understand a little bit about yourself. It's surprising how many people miss this first stage out, and then wonder why they can't decide what to do or don't feel suited to what they end up doing. However, it's really not that complicated; you just need to reflect on a number of questions. Questions such as:

- What is most important to you?
- What are your core values?
- What do you want and need from a job?
- What motivates and interests you?
- What strengths, skills, and experience do you have?

It is answers to these questions, and questions like them, that will help you to know yourself. And knowing yourself will help you to decide the kind of thing you want to do. Do you want to be a research scientist in a pharmaceutical company? An academic? A consultant in a professional services firm? A journalist? A business developer in a start-up organization? Your own boss? The possibilities are endless! It does take a bit of time and effort (which is why it's so often missed out), but it is a vital starting point. Try talking the questions over with people who know you well, such as your friends or family—not that you have to do what they think you should do, but it will provide a useful perspective.

Employers want you to know yourself, too; they are trying to attract not just bright, competent graduates, but bright, competent graduates who have an interest in and passion for their business or profession. Employers spend significant amounts of time and money looking for, and then training their graduates; they don't want to recruit someone who, three months into the job, decides they don't want to be an accountant after all and want to do a PhD instead. Therefore, their selection processes will include elements that try to identify your motivation for wanting to join them (and them in particular, not their sector or profession in general)—if you can't convince them that you are motivated they will filter you out of the process, however bright and competent you may be.

Try this—Knowing yourself

Reflect on what you are like—can you answer some of the questions referred to above:

- What is most important to you?
- What are your core values?
- What do you want and need from a job?
- What motivates and interests you?
- What strengths, skills, and experience do you have?

If you find it difficult to answer these questions, ask someone who knows you really well what their impression of you is in relation to these questions. You may not agree with them, but at least it will get you thinking about them!

15.2.2 Explore your options

Once you understand a bit about yourself and, in particular, what motivates and interests you, you are in a much better position to explore your options. Otherwise it's a bit like walking into a huge library thinking: 'I'd like a book'! You need to have narrowed down your preferences much more in order for your search not to be futile or random (and probably quite demotivating).

A useful starting point is to think about whether you want to use your biosciences degree in a specific or a generic sense. Do you want to become, for example, a researcher, embryologist, botanist, geneticist, or science teacher? Or are you more interested in a graduate job that requires a good degree generally, rather than a biosciences degree in particular? It comes back to thinking about where you want to be and working backwards.

Often, if you want to use your degree in a specific sense, you will need to undertake further study, but you need to be sure of this before you embark upon a higher degree or other form of additional training. You will need to do your research to find out exactly what kind of further study the employers or professions you are interested in require. Do they require a Masters or a PhD, or some other postgraduate qualification? Are there ways in other than the postgraduate qualification route? Is the institution that you are thinking of studying at well regarded by employers? What track record does that institution have of getting its postgraduates into graduate jobs? Postgraduate study is a big investment of time and money, and it's a bad idea to use it as a gap-filler simply because you don't know what you want to do at the end of your undergraduate degree. Otherwise you might find yourself with more letters after your name, a bigger debt, and still no clearer idea as to what you want to do.

If you want to use your degree in a generic sense, this is going to require some research, too. Many, if not most, graduate employers accept students from all disciplines. They are interested in good qualifications, relevant work experience, and transferable skills, but exactly what subject the qualifications are in is usually secondary to the work experience and skills. In addition to thinking about what kind of role you are best suited to (knowing yourself) it is also sensible to think about the current state of the industry or sector you are interested in. For instance, are the police currently recruiting? Is the pharmaceutical industry in the UK growing or shrinking? Are investment banks currently taking on graduates? Are there local companies that are expanding? Will this job exist in five years' time (see Box 15.1)? Knowing the answers to these kinds of questions will give you an indication of how competitive your quest for a graduate job will be. That's not to say you shouldn't aim for the most competitive roles (it's about making yourself as eligible as possible, not random chance), but it's a good idea to have a back-up plan should your first-choice role not come to fruition.

Your university careers service will be able to help you explore your options. If you've never been to see anyone in a careers service before (or even if you have), don't let that put you off! They are there to help you to think through exactly these kinds of issues. If you do some

---

BOX 15.1 Postgraduate study: is it for you?

As we have already indicated, you will already have invested a significant amount of time and money in your current degree programme, and postgraduate training represents further investment. It is important, therefore, that you are very clear in your own mind as to why you are embarking on another qualification. So, the first step is to explore exactly what the employers are looking for.

Postgraduate study can take a number of forms, the most common of which are:

- vocational conversion courses;
- taught Masters programmes;
- postgraduate research degrees.

Vocational conversion courses, as their names suggest, allow you to change direction from your biosciences degree to another professional route. Among the most common such programmes are conversion courses that are linked to specific professions, for example, in law, finance, and areas such as psychology or social work. Some of these programmes may carry academic awards, such as a postgraduate certificate or diploma, or a Masters degree. The most important thing, however, if you are going down this route, is to make sure that the course is recognized by the relevant professional body, and thus allows you an entry into that profession. Your careers service will be able to advise you regarding the range of courses and the different forms of recognition.

Taught Masters programmes are usually one-year courses (if taken full time), which will allow you to gain a higher academic award than your first degree. There is a huge variety of such awards available in a wide range of specialisms. Some may carry bursaries, but these are relatively scarce. The likelihood, therefore, is that you will have to find the money for the course and your living expenses. You need to do your research very carefully, and ask yourself why you are thinking of doing this course and where it could take you. For example, look at the employment profiles of the graduates from the programmes you are considering and, again, talk to your careers service. Just gaining another qualification, for its own sake, is probably not a sensible strategy.

Research degrees also take different forms: at the highest level is the doctorate (PhD or DPhil), which typically lasts three–four years and is predominantly based around a specific research project, although increasingly, they also involve training in a range of research and generic skills. There are also MRes and MPhil opportunities that offer a shorter period of research. The MRes usually allows you to undertake more than one short research project, so you can explore which area you might want to specialize in and, indeed, whether research is for you. There are quite often various awards available to support research students undertaking doctorates, for example, studentships funded by the research councils. Again, you need to consider very carefully why you want to follow this route and you would be strongly advised to gain some practical laboratory experience first, both to strengthen your CV and also to help confirm in your mind that this is what you really want to do.

preparation beforehand, reflecting on specific questions you want answers to or specific things you want help with, it will make the time you spend with an adviser all the more useful.

15.3 Gain experience

At some stage of a selection process for a graduate job (almost any job) you will be asked to respond to statements or questions like:

- Give an example of when you had to manage a challenging situation.
- Describe a situation in which you influenced or motivated people.
- What is your proudest achievement?
- What motivates you?

It's very difficult to provide convincing responses based on academic experience alone. What recruiters will be looking for is a breadth of experience that draws on both curricular and extracurricular activity, not just academic qualifications. Your academic qualifications will only take you so far, and for many roles this is only as far as the first sift of the selection process. The things that will make you stand out are the things that you have in addition to your academic qualifications—your experience.

It's not unusual to hear the following frustration expressed during the search for a job; 'They're asking for work experience—but I can't get work experience without a job and I can't get a job without work experience!' This is understandable, but wrong. There are lots of ways to gain experience apart from employment, particularly in a university context. In fact, there will be far more opportunities available to you than there is time for you to do them; think of freshers' week and all the clubs and societies vying for your attention. There are lots of factors you might consider when deciding what extracurricular activity to get involved in—it might be something that you've never done before that you want to try out; it might be something that you already do that you want to get better at; it might be about meeting new people and making friends. It might be all or none of these reasons, but you should also bear in mind the experience you will gain and the skills you could develop. It doesn't mean you have to be the president of the society or on the committee, but it does mean that you need to be an active member, rather than a passive recipient—someone who gets involved, rather than simply turns up.

Active participation in student clubs and societies is not the only way to gain useful experience while at university: undertaking part-time work, doing an internship, becoming a course rep, learning a language, coaching sports, getting involved in volunteering, a year out in industry, a year abroad, starting your own business, or gaining a skills award are all things that will undoubtedly provide you with useful experience. The trick is picking something that you are motivated to do and that could be useful in terms of where you want to get to. The earlier you get involved in these kinds of things as a student the better; it will give you a chance to try things out and reflect on what you are and are not suited to. As your ideas of what you want to do become clearer and you narrow down your options, you can begin to get involved in things that are more and more relevant to what you want to do. For example, if you want to become a management consultant some commercial experience will be helpful; if you want to become a researcher you will need experience in a lab (see Box 15.2); if you want to get into the media, you can join a student media society.

> **BOX 15.2 How will technology change the job market?**
>
> This is a huge topic and one we cannot possibly do justice to here, but it is important to raise nonetheless. Technological advancements have always changed society and that includes changing work. Periodically, however, changes come along that revolutionize work: the printing press, the steam engine, the mechanization of agriculture, the desktop computer, increasingly sophisticated automation. Just think about how you use the checkout in a supermarket now compared with five years ago. Even more recently, big data, artificial intelligence, and machine learning will all have an increasingly large impact on society and the world of work (think of driverless cars).
>
> A report by consultancy firm PwC in March 2017 ('Will robots steal our jobs? The potential impact of automation on the UK and other major economies') found that up to 30% of existing UK jobs could face automation over the next 15 years. The likelihood of automation, the report says, is highest in the transport, manufacturing, and wholesale and retail sectors, and lowest in education, health, and social work. Every sector will be affected to some degree, however, and while few jobs will entirely disappear, most will be redefined in some way.
>
> The jobs that will be redefined won't just be low-skilled jobs. In their book, *The Future of the Professions* (OUP, 2015), Richard Susskind and Daniel Susskind argue that 'increasingly capable systems' will bring fundamental change in the way that the 'practical expertise' of specialists is made available in society, including doctors, teachers, accountants, architects, consultants, and lawyers.
>
> So what does all this mean for you in terms of making yourself employable? First, it is helpful to understand what jobs are most likely to become automated; these are the jobs that are the most routine and repetitive (hence, the supermarket checkout). Secondly, you can think about what aspects of employment are likely to be least susceptible to automation; these are roles that involve genuine creativity, involve building complex relationships with people, or roles that are highly unpredictable. Thirdly, you can ensure you remain employable by cultivating those skills that are likely to be most in demand (creativity, building complex relationships with people, etc.), so that as the labour market changes, your skills and experience remain well sought after. While this might sound a little daunting, it's worth noting that it has always been the case that those who keep reflecting on and developing their skills are always going to be more competitive in the job market.

15.3.1 Reflect on your skills

'Experience is not what happens to you; it's what you do with what happens to you' is a quote attributed to Aldous Huxley.[1] It makes an important point because it's not just about having experiences that counts, it's learning and developing as a result of those experiences that's important. Again, as with knowing yourself and exploring your options, this will take a bit of time, but it will be time well spent. You need to think about what you are good at and what you're not so good at. You need to think about how to play to your strengths (the things you're good at) and whether the things that you not so good at matter or not in terms of the kind of role that you are looking for—and if they do matter, then you need to think about how you will develop in these areas.

Again, your university careers service should be able to help with this—they can provide you with diagnostic tests to assess your strengths and weaknesses or skills awards programmes that provide a formal structure to help you reflect on and develop your skills.

---

[1] He actually said: 'Experience is not what happens to a man; it is what a man does with what happens to him' (*Texts and Pretexts*, 1932, Chatto & Windus), but this is the given version that has become commonly used.

Try this—Identifying skills developed through problem solving

Ask each one of a group of friends to identify a problem they were faced with and managed to resolve. The problem doesn't just have to be connected to your course; it can be from any area of life. For example, it could be an issue with house mates, something connected to work experience, or any club activities. It may be a problem that affected only you or affected others as well.

As each member of your group describes their problem, make a list of the skills that you think they demonstrated.

15.3.2 Develop your skills

We've been referring to 'skills' throughout this chapter, but have yet to say what these skills actually are. Well, this whole book is about study and communication skills, and certainly those are included, but graduate employers look for more than just these. They do, however, look for a broadly similar set of skills; these may be referred to by different terms (transferable skills, core competencies, skill sets), but they all mean similar things and they all contain similar elements. Typically, graduate employers and employers generally will look for the following skills:

- communication;
- teamwork;
- leadership;
- analysing;
- problem solving and decision making;
- initiative and creativity;
- planning and organizing;
- learning and development;
- self-management;
- using technology.

It can seem quite daunting to be faced with such a broad range of skills, but this is where you need to think about what you are already good at and what you are less good at, and then begin to fill in the gaps step by step. Skills can't be developed in the abstract or in theory; they need to be developed through concrete experience. As we've already identified, opportunities to develop skills through experience abound at university: it's just a case of picking something that motivates and interests you. If it's communications skills that you need to develop, then you need to look for opportunities to present or write; if it's teamwork you need to work on, then put yourself in situations where you have to work in a team to achieve a shared objective; or if it's leadership that you need to evidence, then find opportunities to lead. All of this is obvious, but putting yourself in situations where you are stretched and developed is not necessarily an easy thing to do. However, think where it might get you . . .

Try this—Auditing skills levels

For the following skills associated with group work give yourself a rating and identify an opportunity for further development.

	Novice			Expert	
Do you make a contribution in tutorials, seminars, workshops?	1	2	3	4	5
Can you assume a number of roles, e.g. team leader, team player?	1	2	3	4	5
Do you interact with a range of people to obtain information and assistance?	1	2	3	4	5
Can you give and accept constructive criticism?	1	2	3	4	5

What opportunities might you have in the next few months to practise and develop these skills?

15.3.3 Articulate your skills

Reflecting on and developing your skills are important, but if you can't articulate your skills in a way that employers can relate to and understand the effort may well be wasted. For example, one applicant might describe their part-time bar job on their application form like this:

> Worked part-time in Students' Union bar for 12 months.

Another might describe exactly the same experience like this:

> Committed and reliable member of Students' Union bar staff team; worked as part of a team, providing excellent customer service, often under pressure.

It's not just that one description is shorter and one is longer; the second example actually describes the responsibilities of the role and uses active language to communicate the skills and qualities involved. The candidate in the first description perhaps hasn't reflected on the skills that they've been developing and certainly hasn't articulated them; consequently, they are underselling themselves. Try and get into the habit of regularly reflecting on the skills you are using or developing in these kinds of contexts. If your university encourages personal development planning, then there will be resources to help you do this. Your university careers service will also be able to help you to reflect on the skills you are developing and think about how you can articulate them in the kind of language that employers are looking for.

15.4 Make contact

There are far more new graduates each year than there are graduate jobs, but it's important to realize that the graduate programmes of the big employers represent only a fraction of the graduate jobs available. Small or medium-sized enterprises and business start-ups account for a large proportion of the opportunities available, but are often overlooked because they don't generate as much publicity as the graduate programmes of the big brands. This is important to remember when reading statistics on the average numbers of applicants per graduate job, which vary in any given year, but are usually quite daunting on the face of it. There are three things to bear in mind when reading such figures:

1. the figures relate mainly to the larger employers and competition amongst smaller employers is much less;

2. the volume of applications is different to the quality of applications—just because there are lots of applications (and so the average number of applicants per graduate job is high), it doesn't mean the applications are any good (you need to make good applications, not lots of applications);

3. even in times of very stiff competition some graduate recruiters will not fill all their places—this is because employers will always prefer to carry vacancies, rather than recruit the wrong people.

In a competitive market it's essential that you give yourself as much of an advantage as possible, and that's not just about the skills and experience you have and how you communicate them, but also about the contacts that you make.

15.4.1 Meet employers

One important way that you can stand out to employers is to make them aware of you before you have applied. This might seem unrealistic for the larger employers but it's not necessarily the case; there are many opportunities to meet employers before you even begin the application process, you just need to target them appropriately. Not only is this important to help you stand out, it can also, if approached in the right way, provide you with greater insight into their recruitment and selection process and give you a better idea of what they are looking for in their recruits. This can give you a distinct advantage over other applicants who haven't taken the time to do this.

Your university careers service will be able to tell you about opportunities to meet employers on campus or elsewhere. Some universities are better at attracting employers than others, but all will at least have some employer activity happening. The most common ones are careers fairs and employer presentations, but look out for other opportunities, such as employer visits and special events. Opportunities like these aren't just about graduate roles, but are often the way to find out more about internships and placement years, too.

Before you meet an employer, make sure you do some research—look at their website and find out what you can about their graduate schemes, and their recruitment and selection process; make the most of face-to-face contact by planning questions that you can't find answers to easily online. This will not only give you information that other applicants may not have, but will also show your knowledge of and interest in their organization.

15.4.2 Use your network

In addition to opportunities to meet employers that are put on for you by your university careers service or by individual employers, you should also think about what other contacts you have that you could make use of. Is there anyone you know who already does what you want to do, or is in the industry or profession you are interested in? This could be a direct contact, such as a family member or friend, or an indirect contact, such as a friend-of-a-friend, a member of university staff, or a recent graduate from your department. It might seem tenuous or perhaps a little cheeky, but approaching these kinds of contacts can sometimes make all the difference in your search for information, advice, experience, or even a job.

If you are interested in getting a role in human resources why not contact your university's Human Resources (HR) department and ask to meet with someone to chat about what they do? Or perhaps your friend's brother works for the big retailer you have an interest in working for—why not drop them an email with a few questions? Maybe someone from your department has just finished their PhD and gone on to work in the biotechnology sector—why not contact them via one of your lecturers who knows them? Even busy people will often be happy to respond to these kinds of requests, provided they are presented politely and don't look like they are going to take a large amount of time. It's all about being proactive and doggedly pursuing help in a friendly and respectful fashion. You have nothing to lose and potentially a lot to gain.

Think, too, about how you can create a useful online network. LinkedIn is the most obvious example for professional roles, but other platforms may be appropriate for more niche industries; just look for whatever platforms the companies you are interested in have a presence on. Be careful though if you choose to connect with a company on a platform that you use for social, rather than professional purposes—remember they can see you as well as you seeing them! If using LinkedIn, think about connecting not just with the official company page, but also with people who work in that company. These people could give you potentially useful insights and help you build a professional network. Cold contacts are less likely to be successful so search for alumni from your university who work for the company and personalize the invitation to say you would like to connect. In order to build a network though, you will first need to fill out your profile information; see Section 15.5.1, 'CVs', for help with this (as you can use the information from your CV to populate your profile).

15.4.3 Where to look for jobs

When you are ready to actually look for a vacancy, how do you know where to look (if you are looking for postgraduate courses, see Box 15.3)? It is important to understand that there are advertised vacancies and there are unadvertised vacancies. Estimates vary, but most experts agree that there are more vacancies that are unadvertised than are advertised. Precisely where you look will depend on what it is you are actually looking for and what type of role: voluntary experience, internships, part-time work, or full-time graduate job. Again, this is something that your university careers service can help you with, but here's a brief summary.

Advertised vacancies

- Employers' websites;
- graduate recruitment websites;
- publications, e.g. *The Times Top 100*, *Guardian UK 300*;

BOX 15.3 Postgraduate study: where to look

Before embarking on further study you should have a very clear idea of what it is you want to do and why. There are various forms of further study, including conversion courses, taught postgraduate programmes (Masters programmes, etc.), and research degrees (see Box 15.1).

If you are looking for a taught programme, such as a conversion course or an MSc, then you can search for these online using specific keywords (e.g. 'Law conversion course', or the name of the specific area of Bioscience study, such as 'Molecular Genetics'). Having drawn up an initial list of potential courses, then you should investigate them in the same way that you might have done when choosing your undergraduate programme by looking at:

- the reputation of the institution offering the course;
- the reputation of the course itself and the employment record of the graduates;
- the detailed content of the course;
- the cost of the course and whether there are any funded places;
- the views of the students on the courses offered by the institution—one way of finding this out is through the results of the Postgraduate Taught Experience Survey (PTES); this is rather like the NSS (National Student Survey) for postgraduate students, although at the time of writing the completion rates are much lower;
- the place—is it somewhere where you would like to live?

If you are looking for a PhD, then this is a much more individualized process, which is likely to be focused specifically on the area of research that fascinates you. Opportunities are often advertised online, for example through http://www.findaphD.com or through scientific journals, but often they are obtained through personal contact. During your studies you will have come across areas of research that have fired your interest, and maybe read papers by researchers in the field. At this stage, then, one of the best approaches is to make contact with those researchers, either in your own university or elsewhere, and ask them if they have places for a research student and if there is likely to be any funding. Reputation is important in research and you should look at the websites for the research groups and see what they are publishing and whether they have ongoing funding from research bodies such as the Research Councils (e.g. the Medical Research Council, the Biotechnology and Biological Sciences Research Council, the Natural Environment Research Council) or from major research charities, such as the Wellcome Trust, Cancer Research UK, or Action Research. Talk to the academic staff who are teaching the area you are interested in and see if they can point you in the right direction.

- newspapers—national and local;
- scientific magazines;
- professional associations;
- recruitment agencies;
- company profiles on social media, e.g. LinkedIn, Facebook, Twitter.

Unadvertised vacancies

- *Networking*—via careers fairs, employer campus visits, and social media;
- *personal and professional contacts*;

- *speculative applications*—particularly appropriate for smaller companies;
- *business listings and directories* (local firms)—e.g. yell.com;
- *work experience, internships, placements*—from which you could hear about vacancies;
- *part-time and temporary work*—from which you could hear about vacancies.

15.5 Sell yourself

'Selling yourself' is about articulating your suitability and interest for a role throughout the selection process. Being able to 'sell yourself' is a vital aspect of making yourself employable. You can have a good degree, gained experience, and developed your skills, but if you can't articulate your suitability and interest for the role and the organization you are applying to during the selection process (both formally and informally), then you are unlikely to get very far.

Different employers run their selection processes differently, but there are some broad similarities. Smaller employers typically use a two-stage process involving an application or CV plus covering letter, followed by an interview or interviews. These vacancies are often less visible (possibly unadvertised) and recruit as needed. Larger employers have much more structured processes and their vacancies are usually highly visible in order to compete for the best talent. Larger employers deal with a larger volume of applicants and so will use a multi-stage process to screen applicants. Typically, the stages larger employers use are as follows:

- online registration;
- online application;
- testing (e.g. situational judgement, logical reasoning, numeracy);
- telephone or video interview (sometimes more than one);
- assessment centre;
- interview (often on the same day as the assessment centre and sometimes more than one).

We said earlier that the larger graduate employers may receive hundreds of applications for each position. This highly competitive aspect of the process can leave you feeling somewhat daunted. It's important, however, to be aware of the context: as the graduate recruitment market becomes more competitive, many applicants will reason that they can increase their chances of success by making more applications. This is not necessarily the case. Typically, as the number of applications increases the proportion of quality applications decreases; this is because the more applications people make the less time they are likely to spend tailoring them to the needs of the employer they are applying to. It is much better to make fewer, more focused, better-tailored applications to a small number of targeted employers than to churn out lots of generic, only slightly tweaked applications to a large number of employers, to whom you will struggle to communicate why you want to work for them in particular.

There are many good books and websites available that give information and advice on CVs, applications, and interviews, and your university's careers service will also be able to help you with this (a recurring theme throughout this chapter). What follows, therefore, is simply a summary of the main points to give you an overview; you will need to seek more comprehensive and tailored advice elsewhere.

15.5.1 **CVs**

It's more common for applications for graduate jobs to be via application form, rather than CV. However, a good, up-to-date CV is a really useful document, not least because you can use it to create a professional online profile. We mentioned LinkedIn in 15.4.2, 'Use your network', but in order to create an online network you need to first have a professional online profile. You can use the information from your CV to populate your online profile (or vice versa, if you've already got a LinkedIn profile, but not a CV). Even if you don't use a CV to apply for graduate jobs you will probably use it to apply for part-time or temporary work. Importantly, it will also:

- help you reflect on your strengths and areas for development;
- form the basis for your applications;
- act as a useful aide memoire before interviews or assessment centres to remind you of your qualifications, experience, and skills.

These are just the benefits of a CV from your perspective, but what about the audience for whom you are writing—what are employers looking for? A key fact that you need to bear in mind when writing a CV is that employers or recruiters who accept CVs get a lot of them and so they can't spend very long reading them. A good CV, therefore, needs to communicate relevant information clearly and succinctly in terms of both structure and content.

Structure

The well-structured CV will help the reader see relevant information quickly and easily, in order to make a quick judgement as to whether or not they want to read it in more detail. A badly written CV won't even get past this first look. The information needs to be laid out in a clear and accessible way that helps the reader see the information they are looking for. Typically, a graduate-level CV would include the following headings:

- personal details (name, email, phone number, etc.);
- a brief summary/profile statement (to encourage the reader to read more—highlight your key strengths and make sure you back them up in the rest of the CV);
- education (details of your education to date to an appropriate level, including fuller details about courses relevant to the job—GCSEs would normally be abbreviated to the number and grades, e.g. 6 As (including Maths and English) and 3 Bs);
- work experience (put your most recent and/or relevant experience first);
- skills (you can put these as part of the work experience or include them under their own heading—make sure you know what kind of skills the employer is looking for and then evidence them accordingly, e.g. communication, teamwork, IT, etc.);

- interests (specify your level of involvement and, if possible, use examples that have some relevance to the job);
- references, which can just be 'available on request' (usually one academic and one from a work situation—make sure you ask their permission first and the job description with them so they know what it is you're applying for).

Content

The content of your CV must be tailored to the role and organization you are applying to. You will have baseline content that you can use as a starting point, but it is vital to adapt the content to the role; think carefully about your experience and skills in the context of what is being asked for. Different people have different tendencies; some will tend to oversell themselves and some will tend to undersell themselves. If you are inclined to oversell, then double-check that you have strong evidence to back up your claims (and certainly don't make things up—this almost always causes problems later); if you are inclined to undersell, now is not the time for modesty! Regardless of where you sit on this spectrum, everyone will benefit from a second opinion, so get comments from people who know you well (family and friends) and people who can offer expertise (a contact who knows the organization or job, or your university careers service). If you find yourself really struggling it might be simply because you don't have the qualifications, skills, or experience that are required; this is why it's important to start early (see Section 15.2, 'Engage early'), so that you use the time available to you during your degree to make yourself as eligible as possible when you graduate.

> **Try this—Applying for a graduate position**
>
> Find an advert for a graduate job that interests you (your careers service can help you with where to look). Analyse the skills and experience that you think are required for the position and produce an appropriate CV, then get some feedback from your careers service on it.

15.5.2 Covering letters

The purpose of the covering letter is to personalize your approach to an employer so demonstrating that you have taken time to research the job vacancy and the organization, and put some real thought into your application. Your covering letter also provides you with another opportunity to establish your suitability for the job and to emphasize your key 'selling points'. Typically, a covering letter would include the following sections or paragraphs:

- where your saw the role advertised or why you are making a speculative approach;
- what it is that interests you about the role and the company, and how you have come to this conclusion (this is where to put in any contact that you have had with the company, however brief);

- why you are suitable for the role and the organization (highlight your key selling points—your qualifications, skills, and experience—and match these against the person specification for the job you are applying for);

- a closing sentence about your availability and that you are looking forward to hearing from them.

In all of these, you will need to convey a genuine interest in and enthusiasm for the role.

Try this—Preparing a covering letter

Prepare a covering letter to accompany the CV you created for the previous 'Try this . . .' activity.

15.5.3 Application forms

As mentioned earlier, a good, up-to-date CV will form the starting point for filling in an application form. Qualifications and work experience can probably just be lifted from your CV and dropped into the relevant sections of an application form; however, the rest of the form will take much more thought and effort. As with CVs, employers will receive hundreds, possibly thousands of applications, so first impressions really count. Make sure that you have researched the role and company thoroughly, and that you fulfil the minimum criteria they are asking for. After basic information, like personal details, qualifications, and experience, the application form will usually consist of a series of focused questions designed to ascertain specific information. Some questions will require you to provide credible evidence to back up your claims of specific skills and strengths (try using the STAR model to structure your answers: Situation, Task, Action, Result—see Box 15.4); other questions will attempt to establish your motivations for applying, and your suitability for the role and organization. Just like an essay, a good application takes a long time to research and write, so make sure you give yourself plenty of time. As with CVs, second opinions are really helpful, both from people who can comment on your skills and people who can advise you on the role or organization, so make sure you seek out help.

15.5.4 Testing

As part of the selection process many employers use aptitude tests and personality questionnaires to assess suitability. Such tests are structured ways of evaluating how people perform on various tasks or react to different circumstances.

Aptitude tests

These aim to measure your competencies and intellectual capabilities, as well as your logical and analytical reasoning abilities in a particular area. They aim to assess your abilities to use specific job-related skills and to predict subsequent job performance. The most commonly used tests are:

- verbal reasoning;
- numerical reasoning;
- situational judgement.

BOX 15.4 The STAR technique

The STAR technique provides a framework to help you answer competency-based questions on application forms or in interviews. Competency-based questions are questions are usually focused on a particular skill and you are asked to provide evidence that you have performed it well in the past (albeit, probably in a different setting). Examples include:

- Tell us about a time you had to adjust your communication approach to suit a particular audience.
- Give an example of a time when you had to make a difficult decision.
- Describe a situation in which you have worked as part of a team. What contribution did you make?

The framework is:

- *Situation*—briefly set the scene (where and when?)
- *Task*—what was the aim and what was your role?
- *Action*—what did *you* do and how did *you* contribute (this is the main part of the answer)?
- Result—what results were achieved (you will want to be positive here but it's also good to say how the result could have been improved)?

Example
Question: Tell us about a time you had to adjust your communication approach to suit a particular audience.

Answer:
- (*Situation*) I'm a student ambassador for my university and that means that I occasionally speak to local schools about what it's like to study at university.
- (*Task*) I was asked to talk to a group of year 10 students about what it's like to study at university.
- (*Action*) I was given the slides that had been used previously, but when I looked at them I decided that I wanted to do things a bit differently. I knew I was only going to be talking to a group of about five students, so a PowerPoint presentation seemed a bit over the top. I wanted to make it less formal and encourage more interaction. So, I decided to run a much less formal question-and-answer session, as I thought they would probably have questions from the larger group presentation that they would be attending before mine.
- (*Result*) I'm pleased to say it worked really well! Four of the five students asked questions and I seemed to strike up quite a rapport. Perhaps next time I might ask them to write a question each to encourage the quieter ones a bit, but other than that I'd probably do the same again.

Aptitude tests, as with any part of the selection process, can be quite nerve-wracking, but perhaps particularly so because of their unfamiliarity. The best way to increase both your confidence and performance is to practise. Ask your university careers service for help and for links to online practice tests.

Personality questionnaires

These explore the way in which you do things and how you behave in certain circumstances. They can assess your preferences, motivations, interests, values, and attitudes. You should give careful thought to your responses, but for personality questionnaires there are no right or wrong answers (hence, it being called a 'questionnaire', rather than a 'test'). It may be tempting to try to respond to personality questionnaires in ways that you think will attract the prospective employer, rather than necessarily reflecting your actual personality traits. This is not such a good idea as the questionnaire may (certainly the more sophisticated ones will) include questions that evaluate internal consistency across the questionnaire as a whole, and this is likely to pick up on any areas where you are creating an unrealistic portrayal of yourself.

For both aptitude tests and personality questionnaires, make sure that you give yourself time and space, and a quiet, distraction-free environment to complete them in one sitting.

15.5.5 Telephone or video interview

As we mentioned earlier, larger employers use multi-stage selection processes to deal with large numbers of applicants; this often includes a telephone (or perhaps more commonly now, video) interview after the testing, but before the assessment centre or face-to-face interview stages.

Telephone interview

A telephone interview is usually pre-arranged so you know when to expect it, but can be unannounced; either way, some preparation is vital, just as it is for a face-to-face interview.

Assuming the telephone interview is prearranged you can *prepare your environment* accordingly to ensure that you have access to relevant information and that the space in which you take the call is quiet and free from distractions. Make sure you have a copy of your application form and CV to hand (even if you haven't submitted a CV, it's a useful summary of key points) and let the people you live with know that you are expecting a call and so don't want to be disturbed. If you get an unannounced call and you don't have relevant documents to hand or the environment is too noisy or distracting, then ask if you can arrange a more suitable time to take the call—you are not obliged to speak there and then if the call was unexpected.

In addition to preparing your environment you should also *prepare for the interview* itself. Preparing for telephone interviews is very similar, however, to preparing for face-to-face interviews, so see the *Interviews* Section in 15.5.7 for more information on this.

During the interview itself you will need to be alert and well prepared. Telephone interviews are different in several ways to face-to-face interviews and you need to be aware of these differences in order to perform well. The biggest difference is that there are no visual clues for either you (the interviewee) or them (the interviewer); this means that the tone and rhythm of your voice become much more important. If you speak in a monotone voice this will be more of a problem on the telephone as there will be no facial expressions to help the interviewer interpret the meaning of, or the emotions behind, what you are saying. The following tips will help:

- *Smile!* Silly as it might sound, smiling while you talk really helps. You will come across as more friendly and confident. Use gestures as in normal conversation and be enthusiastic where appropriate.
- *Be organized*—have a pen and paper to hand, and your diary ready in case they like the sound of you and want to invite you to a meeting.
- *Try not to be put off by pauses from the interviewer*—they may be taking notes (this happens in face-to-face interviews, too).

Remember, a telephone interview is a precursor to a face-to-face interview—it is important to create a good impression and answer the questions as best you can. Don't fall into the trap of under-preparing for a telephone interview because it seems less formal or less important than a face-to-face interview—it is an important stage of the selection process and can be the point at which you are excluded from the process or the point at which you are flagged up as a strong potential candidate.

Video interview

A video interview at this stage in the process is not simply a remote alternative to the face-to-face interview (see the *Interviews* section in 15.5.7) but rather involves the interviewee being recorded responding to pre-set questions, and it is not conducted in real time. Candidates are invited to an 'interview' via a web link and are then invited to respond to pre-set questions, which will be reviewed at a later stage by an assessor. Employers are increasingly using this format instead of telephone interviews as it provides the added visual element, can be easily reviewed, and (importantly) doesn't require the assessor to be available at the same time as the candidate.

Once you are signed in to the video interview platform the format is typically: you are asked a question; you are given a minute or so to prepare your answer; the software then records you responding to the question; you then move on to the next question. The precise details will depend on how the system has been configured, so practice is important. Your university careers service will be able to provide help, which often will include an opportunity to practice video-based interviews (and play them back to yourself so you can assess your own performance) and even provide a bookable room in which to have your interview.

Much of the advice detailed above for telephone interviews applies, but in addition you should also consider the following:

- *Prepare your environment*—clearly a video interview allows you to be seen as well as heard, so think about how the 'set' looks (Is it well lit? Is it tidy? Is it quiet?). Check the technology is working beforehand, and if you are unsure about using your own room then try and book a room in you careers service).

- *Prepare for the interview*—in addition to preparing for likely questions (see Section 15.5.7, 'Interviews') you should also dress appropriately. What 'appropriate' means will depend on the context, but it's always better to be slightly too smart, than not smart enough (you won't need to polish your shoes, but dressing the part will probably help).

- *During the interview*—the software will probably allow you to practice your answers and play them back so you can review them, so make sure you take account of this. Use body language appropriately, and don't forget to try and smile!

15.5.6 Assessment centres

Many employers use extended selection exercises as part of their recruitment process; these are often run as an *assessment centre*. Assessment centres are one of the most reliable indicators of future job performance and usually involve a series of exercises designed to show the selectors whether or not you possess the skills required. Typically assessment centres include the following elements (assuming testing—Section 15.5.4—has already been part of the selection process):

- company presentation;
- group tasks;
- individual tasks;
- interview.

Company presentation

The *company presentation* is usually at the beginning of the day, immediately following an introduction and welcome. You should have researched the company well already as part of the previous stages of the selection process, but it is important to remind yourself of what you have already discovered. During the presentation you need to pay close attention to what is being said and perhaps think of a question that you would like to ask at the end, should you be invited to do so. The key points made during the presentation will probably be useful to you for other aspects of the day, so make sure you are concentrating. You may find it helpful to take *brief* notes, but don't let taking notes distract you from what is being said.

Group tasks

The type of *group tasks* used will vary considerably, depending on what the assessors are trying to assess, but it will usually be a discussion, a role play, or a business game. In a *discussion* groups are given a topic to discuss and will be expected to agree a response and present back their conclusions. Assessors will observe candidates' interpersonal skills and the contribution everyone makes to the group. In a *role play* you could be asked to play a role in a simulated exercise. The assessor will look at how you put your views across, whether you can construct a convincing argument and how well you influence and persuade others. A *business game* will involve tackling a simulated business situation, where a task has to be performed, targets met, and difficulties overcome.

These exercises often feel a bit artificial and can make you feel quite self-conscious, but it is important to try to involve yourself as fully as possible and make a positive contribution to the group. It's not about talking the most or being dominant, neither is it about hanging back and expecting everyone else to do the work; it's about putting forward your ideas in a convincing and constructive fashion, while encouraging other people to do the same.

Individual tasks

These might take the form of an in-tray (or 'e-tray') exercise that might involve reading and prioritizing information, identifying action required and any deadlines, drafting replies, preparing reports, and other actions. Alternatively, you might be asked to give a presentation—you may be given the topic in advance (so you can prepare) or you might be asked to prepare it on the day. Regardless of whether you have advanced warning or not, the key to success lies in anticipating your audience's needs, delivering within the specified time frame, and communicating clearly and concisely.

Interview

Finally, you will need to be prepared for an *interview* as part of any assessment centre. Even if you had a telephone or first-stage interview previously you need to expect more in-depth questioning. Your knowledge, your ability to do the job, and your motivation will be under close scrutiny. For information on preparing for interviews see section 15.5.7, 'Interviews'.

Throughout the assessment centre you should assume that every aspect is assessed. You will need to arrive in good time, be polite to everyone, and join in, even at meal times. If you find small talk difficult ask other candidates about their courses or the assessors about how long they have been with the company. When presented with new material for exercises take time to make sure you understand the key information and the requirements of the task. Try to stay calm and focused throughout; don't dwell on any mistakes, just concentrate on doing well in the next task. After the assessment centre, make notes on your experience and ask for feedback on your performance.

15.5.7 Interviews

Interviews are usually the final stage of the selection process and the culmination of all your hard work up to this point. Without giving a good interview you won't get the job, but in order to get an interview you will have needed to have shown suitability to and potential for the role. Interviews are always nerve-wracking experiences, but you can use the nerves to your advantage to help you stay sharp and focused. You also need to stay positive—the organization must have some confidence in you for you to have got to this stage and you need to continue to give them reasons for that confidence.

There are many books and websites giving advice about how to perform well at interviews. The main points though are to do your research, to practise answers to possible questions, and to prepare some questions to ask.

You will have had to have done some *research* in order to get to the interview stage of any selection process and it is important that this information is fresh in your mind at interview. It can be easy to think you have done your research already because you did that at the beginning of the selection process, but you will need to do more prior to the interview itself. The company's website is the place to start when researching a company, but stories about the company in the news (try a Google search for the company and then choose the 'News' results), plus information you can glean from meeting face-to-face with them (see Section 15.4.1, 'Meet employers') are also vital sources of information.

> **Try this—Researching a company**
>
> Choose a company that you think you might be interested in working for—it could be big or small, well known or not. Do some background research on the company that includes: what they do, how many people they employ, where they are based, whether they have overseas sites, what their newest product or service is.

Most of the interview will be taken up with *answering questions* and you need to allocate your preparation time accordingly. You will need to read and re-read your application, and CV because you will doubtless be asked about some of what you have written at the early stages of the selection process. You also need to have practised answers to questions such as:

- Why have you applied for this role?
- What makes you suitable for this role?
- What do you consider yourself good at doing?
- What are your weaknesses?
- Describe a difficult situation and what you did about it.
- What is your proudest achievement?

You will need not just to think about how you will respond to these and other questions, but also *practise your responses out loud*. Ideally, practise with someone who can give you feedback on how you are coming across and ask you some follow-up questions—a friend, parent, or careers adviser.

The final stage of the interview is usually concerned with *questions you have for them*. It is important not to clam up at this point; try to think of a couple of questions before the interview. Some of your questions may well be answered during the interview, in which case you can say that these have been answered, but it is good to be able to have one or two questions to ask. Ask questions that show you have thought about the job and the company.

Occasionally, interviews are conducted remotely via video conferencing software (e.g. Skype). This isn't common for graduate recruitment (other than earlier in the selection process, as described in Section 15.5.5, 'Telephone or video interview', where specialist video interviewing software is increasingly used), and if you're given a choice we recommend that you attend in person. However, should you find yourself in a video interviewing setting you will need to consider the following:

- Make sure the technology is set up and working well in advance, so you have one less thing to worry about immediately before the interview.
- Ensure that you won't be disturbed; again, your university careers service may allow you to book a room for the purpose.
- Consider the 'set' for the interview—well lit, tidy, and professional is probably best.
- Dress professionally—while this may not seem as important as in a face-to-face setting, dressing the part (top half and bottom half) will help you be more mentally prepared.
- If using your own account (e.g. Skype), rather than a link to the employer's system, make sure you have an appropriate username.

In both your answering and asking questions, in fact throughout the whole selection process, you need to present yourself well—be friendly and be positive. If it doesn't work out, and you are unsuccessful at this or any stage of the selection process, ask for feedback. There will always be room for improvement and, given how much time and effort is involved in completing a selection process, and how much is riding on it, it is important that you use feedback to improve your approach and increase your chances of success.

Chapter summary

Making yourself employable is a vital part of your degree. It is essential that you start thinking about your employability from the beginning of your degree, rather than at the end so that you can enhance your employability as you go along, rather than trying to cram it all in at the end (which doesn't work). While getting a good degree is an important part of finding a good job, it represents only a fraction of what employers look for in potential candidates. Since employers can easily find out about your academic qualifications (they just ask you to tell them what you have got), the selection process is weighted towards finding out the skills, experience, and attributes you can offer an employer in addition to your academic qualifications. Gaining experience, making contact with employers, and selling yourself are essential elements of making yourself more employable.

Index

Page numbers in bold indicate the start of a chapter on the subject.

PUGS
DON'T WEAR PYJAMAS

For Gigi Roz and Ellie the pug. MW

For my lovely boy, Olle. CJ

PUGS
DON'T WEAR PYJAMAS

Michelle Worthington • Cecilia Johansson

NEW FRONTIER PUBLISHING

Tom was looking forward to staying with his Aunt Roz.
His aunt had a new friend for him to play with.
Her name was Ellie.

Tom couldn't believe his eyes.

At bedtime Aunt Roz read Tom a story.

'Pugs don't wear pyjamas,' snorted Tom.
'Ellie does,' smiled Aunt Roz.

The next day they went to the beach for a swim.

'Pugs don't wear hats,' said Tom.

'Ellie does.' Aunt Roz replied.

At lunchtime Tom sat opposite Ellie. 'Pugs don't eat at cafés,' Tom grumbled, shaking his head.

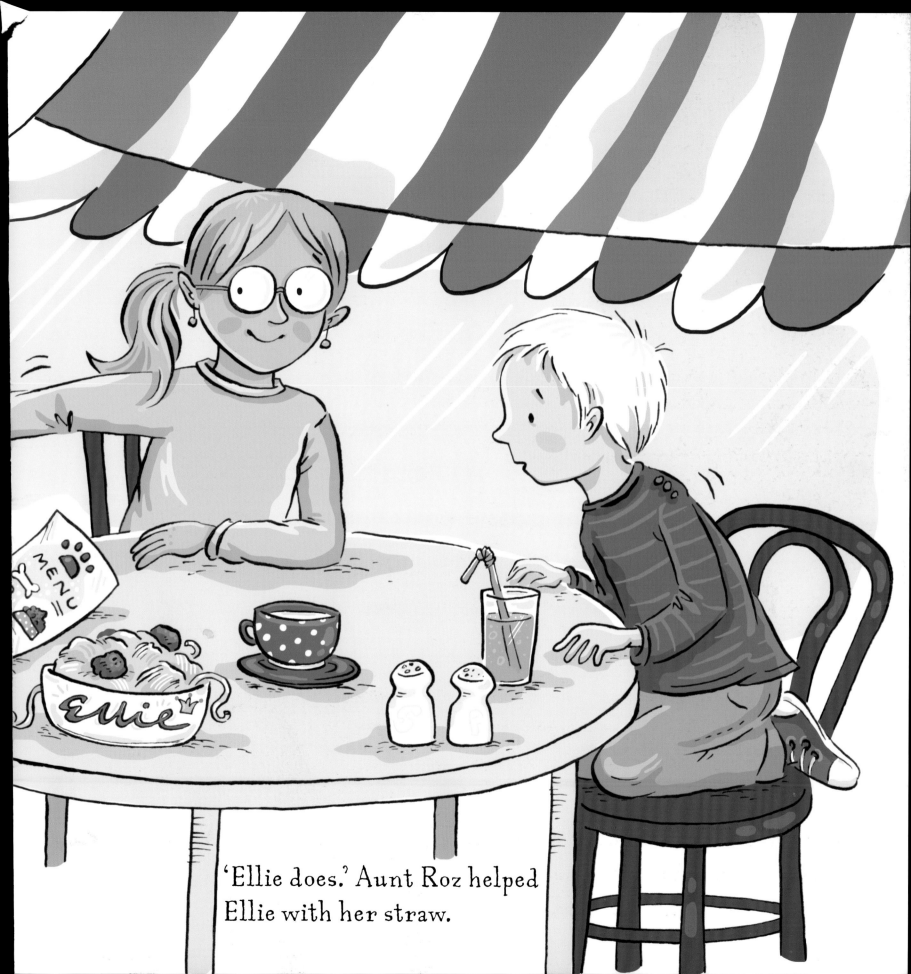

'Ellie does.' Aunt Roz helped
Ellie with her straw.

Everywhere that Tom and his aunt went, Ellie went too.
They went bike riding in the park.

They tried out their new skateboards.

They went down a slide ...

... and climbed a tree.

Ellie made friends wherever she went.

Tom didn't.

Everybody made a fuss of Ellie. Tom stayed quiet.

One afternoon, when Tom was meant to be playing with Ellie, he knocked the latch on the gate.

When Roz called Ellie for her afternoon tea, Ellie didn't come.

She wasn't in the circus tent ...

... and she wasn't behind the shed ...

she wasn't under the chair.

Tom and Aunt Roz looked for Ellie at the beach ...

... the café ...

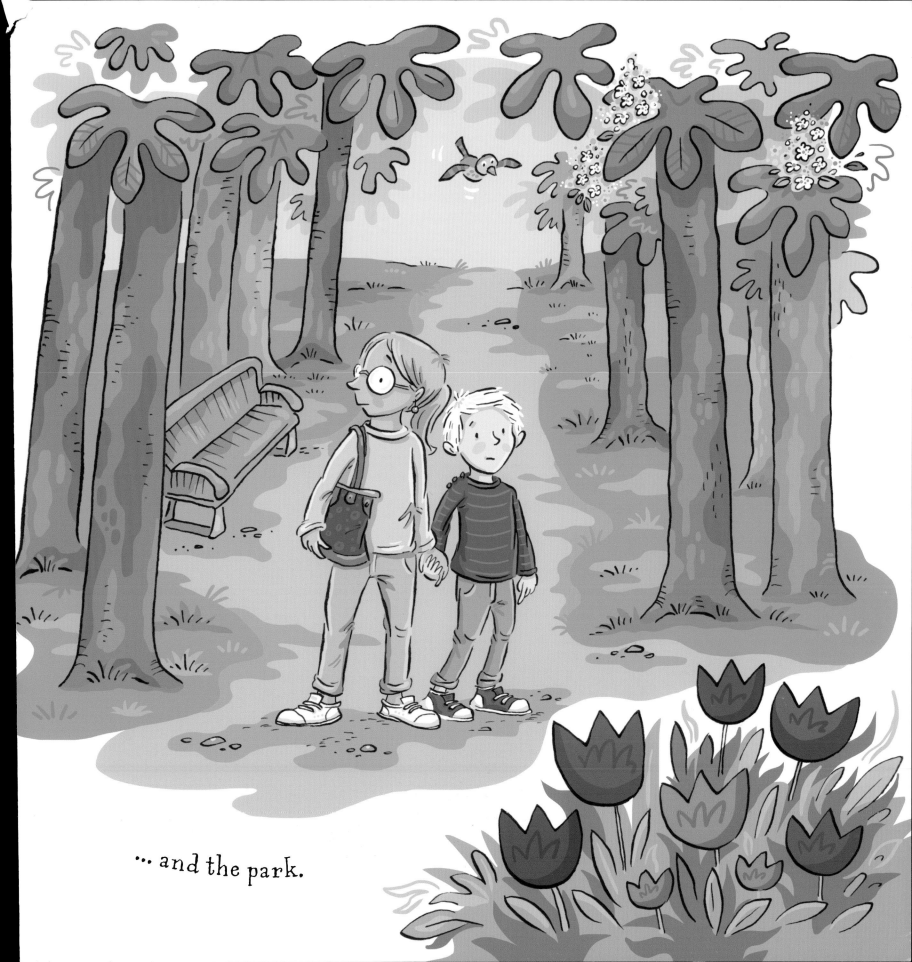

... and the park.

Aunt Roz was worried. Where could Ellie be? Tom felt terrible.

He had promised to help look after her.

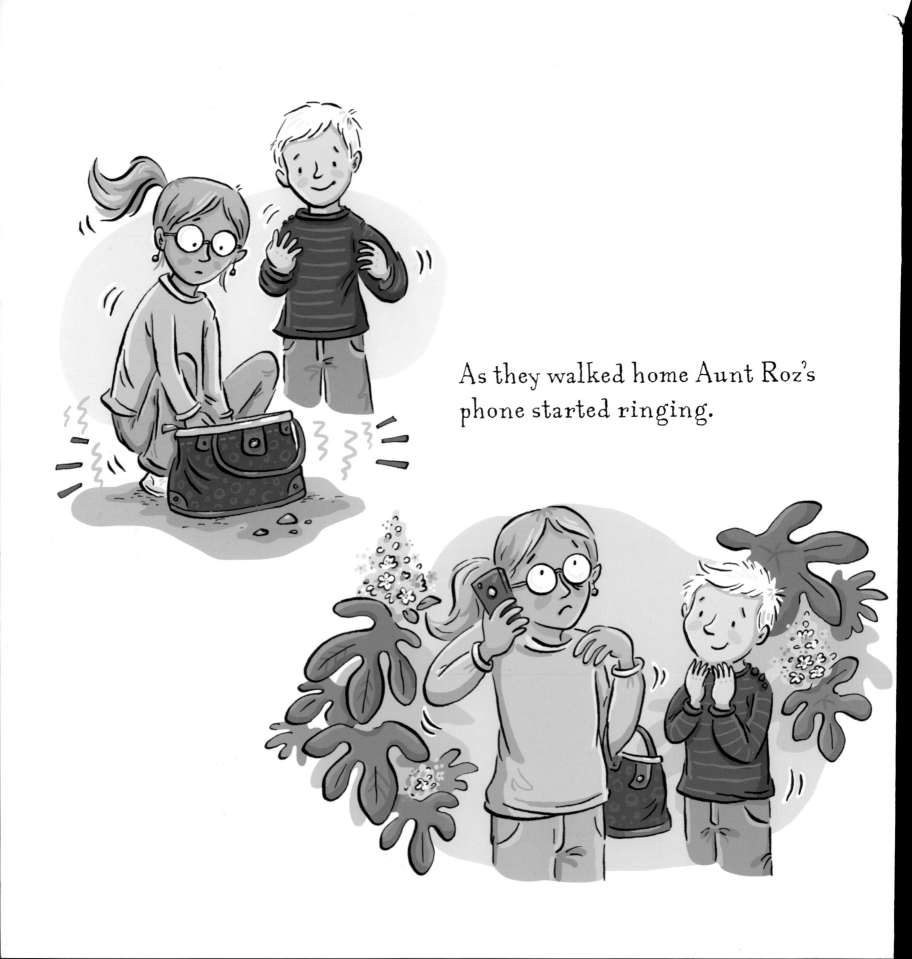

As they walked home Aunt Roz's phone started ringing.

She took Tom's hand and rushed straight to the ...

... pet shop! Ellie had her front paws on the glass.

Aunt Roz scooped Ellie into her arms and gave her a
big fat kiss on her big flat nose.

Tom gave Ellie a pat.

'Ellie wants a guinea pig,' laughed Tom.
'Pugs don't have pets!' snorted the pet shop owner.

'Ellie does,' smiled Tom.